Asa Gray

Gray's lessons in botany and vegetable physiology

Asa Gray

Gray's lessons in botany and vegetable physiology

ISBN/EAN: 9783337278564

Printed in Europe, USA, Canada, Australia, Japan

Cover: Foto ©berggeist007 / pixelio.de

More available books at **www.hansebooks.com**

GRAY'S

LESSONS IN BOTANY

AND

VEGETABLE PHYSIOLOGY,

ILLUSTRATED BY OVER 360 WOOD ENGRAVINGS, FROM ORIGINAL
DRAWINGS, BY ISAAC SPRAGUE.

TO WHICH IS ADDED A COPIOUS

GLOSSARY,

OR

DICTIONARY OF BOTANICAL TERMS.

By ASA GRAY,

FISHER PROFESSOR OF NATURAL HISTORY IN HARVARD UNIVERSITY.

IVISON, BLAKEMAN, TAYLOR & CO.,
NEW YORK AND CHICAGO.
1879.

PREFACE.

THIS book is intended for the use of beginners, and for classes in the common and higher schools, — in which the elements of Botany, one of the most generally interesting of the Natural Sciences, surely ought to be taught, and to be taught correctly, as far as the instruction proceeds. While these Lessons are made as plain and simple as they well can be, all the subjects treated of have been carried far enough to make the book a genuine Grammar of Botany and Vegetable Physiology, and a sufficient introduction to those works in which the plants of a country — especially of our own — are described.

Accordingly, as respects the principles of Botany (including Vegetable Physiology), this work is complete in itself, as a school-book for younger classes, and even for the students of our higher seminaries. For it comprises a pretty full account of the structure, organs, growth, and reproduction of plants, and of their important uses in the scheme of creation, — subjects which certainly ought to be as generally understood by all educated people as the elements of Natural Philosophy or Astronomy are; and which are quite as easy to be learned.

The book is also intended to serve as an introduction to the author's *Manual of the Botany of the Northern United States* (or to any similar work describing the plants of other districts), and to be to it what a grammar and a dictionary are to a classical author. It consequently contains many terms and details which there is no necessity for young students perfectly to understand in the first instance, and still less to commit to memory, but which they will need to refer to as occasions arise, when they come to analyze flowers, and ascertain the names of our wild plants.

To make the book complete in this respect, a full *Glossary, or Dictionary of Terms used in describing Plants*, is added to the volume. This contains very many words which are not used in the *Manual of Botany;* but as they occur in common botanical works, it was thought best to introduce and explain them. All the words in the Glossary which seemed to require it are accented.

It is by no means indispensable for students to go through the volume before commencing with the analysis of plants. When the proper season for botanizing arrives, and when the first twelve Lessons have been gone over, they may take up Lesson XXVIII. and the following ones, and proceed to study the various wild plants they find in blossom, in the manner illustrated in Lesson XXX., &c., — referring to the Glossary, and thence to the pages of the Lessons, as directed, for explanations of the various distinctions and terms they meet with. Their first essays will necessarily be rather tedious, if not difficult; but each successful attempt smooths the way for the next, and soon these technical terms and distinctions will become nearly as familiar as those of ordinary language.

Students who, having mastered this elementary work, wish to extend their acquaintance with Vegetable Anatomy and Physiology, and to consider higher questions about the structure and classification of plants, will be prepared to take up the author's *Botanical Text-Book*, an *Introduction to Structural Botany*, or other more detailed treatises.

No care and expense have been spared upon the illustrations of this volume; which, with one or two exceptions, are all original. They were drawn from nature by Mr. Sprague, the most accurate of living botanical artists, and have been as freely introduced as the size to which it was needful to restrict the volume would warrant.

To append a set of questions to the foot of each page, although not unusual in school-books, seems like a reflection upon the competency or the faithfulness of teachers, who surely ought to have mastered the lesson before they undertake to teach it; nor ought facilities to be afforded for teaching, any more than learning, lessons by rote. A full *analysis of the contents* of the Lessons, however, is very convenient and advantageous. Such an Analysis is here given, in place of the ordinary table of contents. This will direct the teacher and the learner at once to the leading ideas and important points of each Lesson, and serve as a basis to ground proper questions on, if such should be needed.

ASA GRAY.

Harvard University, Cambridge,
        January 1, 1857.

⁎ Revised August, 1868, and alterations made adapting it to the new edition of *Manual*, and to *Field, Forest, and Garden Botany*, to which this work is the proper introduction and companion.

                                                    A. G.

ANALYSIS OF THE LESSONS.*

---

* The numbers in the analysis refer to the paragraphs.

a *

what it depends on: 51. how it becomes incomplete: 51–59. how varied. 53. Definite growth. 54. Indefinite growth. 55. Deliquescent or dissolving stems, how formed. 56. Excurrent stems of spire-shaped trees, how produced. 57. Latent Buds. 58. Adventitious Buds. 59. Accessory or supernumerary Buds. 60. Sorts of Buds recapitulated and defined.

LESSON V. Morphology of Roots. . . . . p. 28.

61–64. Morphology; what the term means, and how applied in Botany. 65. Primary Root, simple; and, 66. multiple. 67. Rootlets; how roots absorb: time for transplantation, &c. 68. Great amount of surface which a plant spreads out, in the air and in the soil; reduced in winter, increased in spring. 69. Absorbing surface of roots increased by the root-hairs. 70. Fibrous roots for absorption. 71. Thickened or fleshy roots as storehouse of food. 72, 73. Their principal forms. 74. Biennial roots; their economy. 75. Perennial thickened roots. 76. Potatoes, &c. are not roots. 77. Secondary Roots, their economy. 78. Sometimes striking in open air, when they are, 79. Aerial Roots; illustrated in Indian Corn, Mangrove, Screw Pine, Banyan, &c. 80. Aerial Rootlets of Ivy. 81. Epiphytes or Air-Plants, illustrated. 82. Parasitic Plants, illustrated by the Mistletoe, Dodder, &c.

LESSON VI. Morphology of Stems and Branches. . . . p. 36.

83–85. Forms of stems and branches above ground. 86. Their direction or habit of growth. 87. Culm, Caudex, &c. 88. Suckers: propagation of plants by division. 89. Stolons: propagation by layering or laying. 90. Offsets. 91. Runners. 92. Tendrils; how plants climb by them: their disk-like tips in the Virginia Creeper. 93. Tendrils are sometimes forms of leaves. 94. Spines or Thorns; their nature: Prickles. 95. Strange forms of stems. 96. Subterranean stems and branches. 97. The Rootstock or Rhizoma, why stem and not root. 98. Why running rootstocks are so troublesome, and so hard to destroy. 99–101. Thickened rootstocks, as depositories of food. 102. Their life and growth. 103. The Tuber. 104. Economy of the Potato-plant. 105. Gradations of tubers into, 106. Corms or solid bulbs: the nature and economy of these, as in Crocus. 107. Gradation of these into, 108. the Bulb: nature of bulbs. 109, 110. Their economy. 111. Their two principal sorts. 112. Bulblets. 113. How the foregoing sorts of stems illustrate what is meant by morphology. 114. They are imitated in some plants above ground. 115. Consolidated forms of vegetation, illustrated by Cactuses, &c. 116. Their economy and adaptation to dry regions.

LESSON VII. Morphology of Leaves. . . . . p. 49.

117. Remarkable states of leaves already noticed. 118, 119. Foliage the natural form of leaves: others are special forms, or transformations; why so called. 120. Leaves as depositories of food, especially the seed-leaves; and, 121. As Bulb-scales. 122. Leaves as Bud-scales. 123. As Spines. 124. As Tendrils. 125. As Pitchers. 126. As Fly-traps. 127–129. The same leaf serving various purposes.

ing. 434. What the living parts of a tree are; their annual renewal. 435. Cambium-layer or zone of growth in the stem; connected with, 436. new rootlets below, and new shoots, buds, and leaves above. 437. Structure of a leaf: its two parts, the woody and the cellular, or, 438. the pulp; this contains the green matter, or Chlorophyll. 439, 440. Arrangement of the cells of green pulp in the leaf, and structure of its epidermis or skin. 441. Upper side only endures the sunshine. 442. Evaporation or exhalation of moisture from the leaves. 443. Stomates or Breathing-pores, their structure and use. 444. Their numbers.

LESSON XXVI. THE PLANT IN ACTION, DOING THE WORK
               OF VEGETATION. . . . . p. 157.

446. The office of plants to produce food for animals. 447. Plants feed upon earth and air. 449. Their chemical composition. 450. Two sorts of material. 451, 452. The earthy or inorganic constituents. 453. The organic constituents. 454. These form the Cellulose, or substance of vegetable tissue; composition of cellulose. 455. The plant's food, from which this is made. 456. Water, furnishing hydrogen and oxygen. 458. Carbonic acid, furnishing, 457. Carbon. 459. The air, containing oxygen and nitrogen; and also, 460. Carbonic acid; 461. which is absorbed by the leaves, 462. and by the roots. 463. Water and carbonic acid the general food of plants. 464. Assimilation the proper work of plants. 465. Takes place in green parts alone, under the light of the sun. 466 – 468. Liberates oxygen gas and produces Cellulose or plant-fabric. 469. Or else Starch; its nature and use. 470. Or Sugar; its nature, &c. The transformations starch, sugar, &c. undergo. 471. Oils, acids, &c. The formation of all these products restores oxygen gas to the air. 472. Therefore plants purify the air for animals. 473. While at the same time they produce all the food and fabric of animals. The latter take all their food ready made from plants. 474. And decompose starch, sugar, oil, &c., giving back their materials to the air again as the food of the plant; at the same time producing animal heat. 475. But the fabric or flesh of animals (fibrine, gelatine, &c.) contains nitrogen. 476. This is derived from plants in the form of Proteine. Its nature and how the plant forms it. 477. Earthy matters in the plant form the earthy part of bones, &c. 478. Dependence of animals upon plants; showing the great object for which plants were created.

LESSON XXVII. PLANT-LIFE. . . . . p. 166.

479. Life; manifested by its effects; viz. its power of transforming matter: 480. And by motion. 481, 482. Plants execute movements as well as animals. 483. Circulation in cells. 484. Free movements of the simplest plants in their forming state. 485. Absorption and conveyance of the sap. 486. Its rise into the leaves. 487. Explained by a mechanical law; Endosmose. 488. Set in action by evaporation from the leaves. 489. These movements controlled by the plant, which directs growth and shapes the fabric by an inherent power. 490 – 492. Special movements of a conspicuous sort; such as seen in the bending, twining, revolving, and coiling of stems and tendrils; in the so-called sleeping and waking states of plants; in movements from irritation, and striking spontaneous motions.

FIRST LESSONS

IN

BOTANY AND VEGETABLE PHYSIOLOGY.

LESSON I.

BOTANY AS A BRANCH OF NATURAL HISTORY.

1. THE subjects of Natural History are, the earth itself and the beings that live upon it.

2. **The Inorganic World, or Mineral Kingdom.** The earth itself, with the air that surrounds it, and all things naturally belonging to them which are destitute of life, make up the mineral kingdom, or inorganic world. These are called *inorganic*, or unorganized, because they are not composed of *organs*, that is, of parts which answer to one another, and make up a whole, such as is a horse, a bird, or a plant. They were formed, but they did not grow, nor proceed from previous bodies like themselves, nor have they the power of producing other similar bodies, that is, of reproducing their kind. On the other hand, the various living things, or those which have possessed life, compose

3. **The Organic World,** — the world of organized beings. These consist of *organs;* of parts which go to make up an *individual, a being.* And each individual owes its existence to a preceding one like itself, that is, to a parent. It was not merely formed, but *produced.* At first small and imperfect, it grows and develops by powers of its own; it attains maturity, becomes old, and finally dies. It was formed of inorganic or mineral matter, that is, of earth and air, indeed; but only of this matter under the influence of life: and after life departs, sooner or later, it is decomposed into earth and air again.

1

4. The organic world consists of two kinds of beings; namely, 1. *Plants* or *Vegetables*, which make up what is called the *Vegetable Kingdom*; and, 2. *Animals*, which compose the *Animal Kingdom*.

5. **The Differences between Plants and Animals** seem at first sight so obvious and so great, that it would appear more natural to inquire how they resemble rather than how they differ from each other. What likeness does the cow bear to the grass it feeds upon? The one moves freely from place to place, in obedience to its own will, as its wants or convenience require: the other is fixed to the spot of earth where it grew, manifests no will, and makes no movements that are apparent to ordinary observation. The one takes its food into an internal cavity (the stomach), from which it is absorbed into the system: the other absorbs its food directly by its surface, by its roots, leaves, &c. Both possess organs; but the limbs or members of the animal do not at all resemble the roots, leaves, blossoms, &c. of the plant. All these distinctions, however, gradually disappear, as we come to the lower kinds of plants and the lower animals. Many animals (such as barnacles, coral-animals, and polyps) are fixed to some support as completely as the plant is to the soil; while many plants are not fixed, and some move from place to place by powers of their own. All animals move some of their parts freely; yet in the extent and rapidity of the motion many of them are surpassed by the common Sensitive Plant, by the Venus's Fly-trap, and by some other vegetables; while whole tribes of aquatic plants are so freely and briskly locomotive, that they have until lately been taken for animals. It is among these microscopic tribes that the animal and vegetable kingdoms most nearly approach each other, — so nearly, that it is still uncertain where to draw the line between them.

6. Since the difficulty of distinguishing between animals and plants occurs only, or mainly, in those forms which from their minuteness are beyond ordinary observation, we need not further concern ourselves with the question here. One, and probably the most absolute, difference, however, ought to be mentioned at the outset, because it enables us to see what plants are made for. It is this: —

7. Vegetables are nourished by the mineral kingdom, that is, by the ground and the air, which supply all they need, and which they are adapted to live upon; while animals are entirely nourished by vegetables. The great use of plants therefore is, to take portions of

earth and air, upon which animals cannot subsist at all, and to convert these into something upon which animals can subsist, that is,
into food. *All food is produced by plants.* How this is done, it is
the province of Vegetable Physiology to explain.

8. **Botany** is the name of the science of the vegetable kingdom in
general.

9. **Physiology** is the study of the way a living being lives, and
grows, and performs its various operations. ' The study of plants in
this view is the province of *Vegetable Physiology.* The study of the
form and structure of the organs or parts of the vegetable, by which
its operations are performed, is the province of *Structural Botany.*
The two together constitute *Physiological Botany.* With this department the study of Botany should begin; both because it lies
at the foundation of all the rest, and because it gives that kind of
knowledge of plants which it is desirable every one should possess;
that is, some knowledge of the way in which plants live, grow, and
fulfil the purposes of their existence. To this subject, accordingly,
a large portion of the following Lessons is devoted.

10. The study of plants as to their *kinds* is the province of *Systematic Botany.* An enumeration of the kinds of vegetables, as far
as known, classified according to their various degrees of resemblance
or difference, constitutes a general *System of plants.* A similar account of the vegetables of any particular country or district is called
a *Flora* of that country or district.

11. Other departments of Botany come to view when — instead
of regarding plants as to what they are in themselves, or as to their
relationship with each other — we consider them in their relations
to other things. Their relation to the earth, for instance, as respects
their distribution over its surface, gives rise to *Geographical Botany,*
or *Botanical Geography.* The study of the vegetation of former
times, in their fossil remains entombed in the crust of the earth,
gives rise to *Fossil Botany.* The study of plants in respect to their
uses to man is the province of *Agricultural Botany, Medical Botany,*
and the like.

LESSON II.

THE GROWTH OF THE PLANT FROM THE SEED.

12. The Course of Vegetation. We see plants growing from the seed in spring-time, and gradually developing their parts : at length they blossom, bear fruit, and produce seeds like those from which they grew. Shall we commence the study of the plant with the full-grown herb or tree, adorned with flowers or laden with fruit ? Or shall we commence with the seedling just rising from the ground? On the whole, we may get a clearer idea of the whole life and structure of plants if we begin at the beginning, that is, with the plantlet springing from the seed, and follow it throughout its course of growth. This also agrees best with the season in which the study of Botany is generally commenced, namely, in the spring of the year, when the growth of plants from the seed can hardly fail to attract attention. Indeed, it is this springing forth of vegetation from seeds and buds, after the rigors of our long winter, — clothing the earth's surface almost at once with a mantle of freshest verdure, — which gives to spring its greatest charm. Even the dullest beholder, the least observant of Nature at other seasons, can then hardly fail to ask : What are plants? How do they live and grow? What do they live upon ? What is the object and use of vegetation in general, and of its particular and wonderfully various forms ? These questions it is the object of the present Lessons to answer, as far as possible, in a simple way.

13. A reflecting as well as observing person, noticing the re-semblances between one plant and another, might go on to inquire whether plants, with all their manifold diversities of form and appearance, are not all constructed on one and the same general plan. It will become apparent, as we proceed, that this is the case ; — that one common plan may be discerned, which each particular plant, whether herb, shrub, or tree, has followed much more closely than would at first view be supposed. The differences, wide as they are, are merely incidental. What is true in a general way of any ordinary vegetable, will be found to be true of all, only with great variation in the details. In the same language, though in varied phrase, the hundred thousand kinds of plants repeat the same

story,— are the living witnesses and illustrations of one and the same plan of Creative Wisdom in the vegetable world. So that the study of any one plant, traced from the seed it springs from round to the seeds it produces, would illustrate the whole subject of vegetable life and growth. It ma'ters little, therefore, what particular plant we begin with.

14. **The Germinating Plantlet.** Take for example a seedling Maple. Sugar Maples may be found in abundance in many places, starting from the seed (i. e. *germinating*) in early spring, and Red Maples at the beginning of summer, shortly after the fruits of the season have ripened and fallen to the ground. A pair of narrow green leaves raised on a tiny stem make up the whole plant at its first appearance (Fig. 4). Soon a root appears at the lower end of this stemlet ; then a little bud at its upper end, between the pair of leaves, which soon grows into a second joint or stem bearing another pair of leaves, resembling the ordinary leaves of the Red Maple, which the first did not. Figures 5 and 6 represent these steps in the growth.

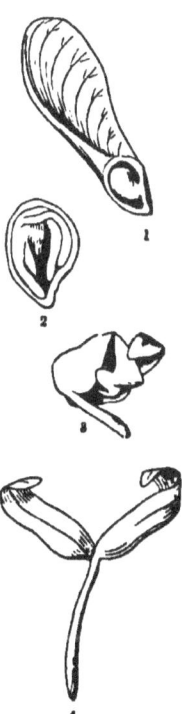

15. Was this plantlet formed in the seed at the time of germination, something as the chick is formed in the egg during the process of incubation ? Or did it exist before in the seed, ready formed ? To decide this question, we have only to inspect a sound seed, which in this instance requires no microscope, nor any other instrument than a sharp knife, by which the coats of the seed (previously soaked in water, if dry) may be laid open. We find within the seed, in this case, the little plantlet ready formed, and nothing else (Fig. 2) ;— namely, a pair of leaves like those of the earliest seedling (Fig. 4), only smaller, borne on a stemlet just like that of the seedling, only much shorter, and all snugly coiled up within the protecting seed-coat. The plant then exists beforehand in the seed, in miniature. It was not formed, but only devel-

FIG. 1. A winged fruit of Red Maple, with the seed-bearing portion cut open, to show the seed. 2. This seed cut open to show the embryo plantlet within, enlarged. 3. The embryo taken out whole, and partly unfolded. 4. The same after it has begun to grow ; of the natural size.

1 *

oped, in germination ; when it had merely to unfold and grow, — to elongate its rudimentary stem, which takes at the same time an upright position, so as to bring the leaf-bearing end into the light and air, where the two leaves expand ; while from the opposite end, now pushed farther downwards into the soil, the root begins to grow. All this is true in the main of all plants that spring from real seeds, although with great diversity in the particulars. At least, there is hardly an exception to the fact, that *the plantlet exists ready formed in the seed*, in some shape or other.

16. The rudimentary plantlet contained in the seed is called an *Embryo*. Its little stem is named the *Radicle*, because it was supposed to be the root, when the difference between the root and stem was not so well known as now. It were better to name it the *Caulicle* (i. e. little stem) ; but it is not expedient to change old names. The seed-leaves it bears on its summit (here two in number) are technically called *Cotylèdons*. The little bud of undeveloped leaves which is to be found between the cotyledons before germination in many cases (as in the Pea, Bean, Fig. 17, &c.), has been named the *Plumule*.

17. In the Maple (Fig. 4), as also in the Morning-Glory (Fig. 28), and the like, this bud, or plumule, is not seen for some days after the seed-leaves are expanded. But soon it appears, in the Maple as a pair of minute leaves (Fig. 5), erelong raised on a stalk which carries them up to some distance above the cotyledons. The plantlet (Fig. 6) now consists, above ground, of two pairs of leaves, viz. : 1. the cotyledons or seed-leaves, borne on the summit of the original stemlet (the radicle) ; and 2. a pair of ordinary leaves, raised on a second joint of stem which has grown from the top of the first. Later, a third pair of leaves is formed, and raised on a third joint of stem, proceeding from the summit of the second (Fig. 7), just as that did from the first; and so on, until the germinating plantlet becomes a tree.

FIG. 5. Germinating Red Maple, which has produced its root beneath, and is developing a second pair of leaves above. 6. Same, further advanced.

18. So the youngest seedling, and even the embryo in the seed is already an epitome of the herb or tree. It has a stem, from the lower end of which it strikes root : and it has leaves. The tree itself in its whole vegetation has nothing more in kind. To become a tree, the plantlet has only to repeat itself upwardly by producing more similar parts, — that is, new portions of stem, with new and larger leaves, in succession, — while beneath, it pushes its root deeper and deeper into the soil.

19. **The Opposite Growth of Root and Stem** began at the beginning of germination, and it continues through the whole life of the plant. While yet buried in the soil, and perhaps in total darkness, as soon as it begins to grow, the stem end of the embryo points towards the light, — curving or turning quite round if it happens to lie in some other direction, — and stretches upwards into the free air and sunshine ; while the root end as uniformly avoids the light, bends in the opposite direction to do so if necessary, and ever seeks to bury itself more and more in the earth's bosom. How the plantlet makes these movements we cannot explain. But the object of this instinct is obvious. It places the plant from the first in the proper position, with its roots in the moist soil, from which they are to absorb nourishment, and its leaves in the light and air, where alone they can fulfil their office of digesting what the roots absorb.

20. So the seedling plantlet finds itself provided with all the *organs of vegetation* that even the oldest plant possesses, — namely, root, stem. and leaves ; and has these placed in the situation where each is to act, — the root in the soil, the foliage in the light and air. Thus established, the plantlet has only to set about its proper work.

21. **The different Mode of Growth of Root and Stem** may also be here mentioned. Each grows, not only in a different direction, but in a different way. The stem grows by producing a set of joints, each from

FIG. 7. Germinating Red Maple, further developed.

the summit of its predecessor; and each joint elongates throughout every part, until it reaches its full length. The root is not composed of joints, and it lengthens only at the end. The stem in the embryo (viz. the radicle) has a certain length to begin with. In the pumpkin-seed, for instance (Fig. 9), it is less than an eighth of an inch long: but it grows in a few days to the length of one or two inches (Fig. 10), or still more, if the seed were deeper covered by the soil. It is by this elongation that the seed-leaves are raised out of the soil, so as to expand in the light and air. The length they acquire varies with the depth of the covering. When large and strong seeds are too deeply buried, the stemlet sometimes grows to the length of several inches in the endeavor to bring the seed-leaves to the surface. The lengthening of the succeeding joints of the stem serves to separate the leaves, or pairs of leaves, from one another, and to expose them more fully to the light.

22. The root, on the other hand, begins by a new formation at the base of the embryo stem; and it continues to increase in length solely by additions to the extremity, the parts once formed scarcely elongating at all afterwards. This mode of growth is well adapted to the circumstances in which roots are placed, leaving every part undisturbed in the soil where it was formed, while the ever-advancing points readily insinuate themselves into the crevices or looser portions of the soil, or pass around the surface of solid obstacles.

8

LESSON III.

GROWTH OF THE PLANT FROM THE SEED. — *Continued.*

23. So a plant consists of two parts, growing in a different manner. as well as in opposite directions. One part, the root, grows downwards into the soil: it may, therefore, be called the *descending axis.* The other grows upwards into the light and air: it may be called the *ascending axis.* The root grows on continuously from the extremity, and so does not consist of joints, nor does it bear leaves, or anything of the kind. The stem grows by a succession of joints, each bearing one or more leaves on its summit. Root on the one hand, and stem with its foliage on the other, make up the whole plantlet as it springs from the seed; and the full-grown herb, shrub, or tree has nothing more in kind, — only more in size and number. Before we trace the plantlet into the herb or tree, some other cases of the growth of the plantlet from the seed should be studied, that we may observe how the same plan is worked out under a variety of forms, with certain differences in the details. The materials for this study are always at hand. We have only to notice what takes place all around us in spring, or to plant some common seeds in pots, keep them warm and moist, and watch their germination.

24. **The Germinating Plantlet feeds on Nourishment provided beforehand.** The embryo so snugly ensconced in the seed of the Maple (Fig. 2, 3, 4) has from the first a miniature stem, and a pair of leaves already green, or which become green as soon as brought to the light. It has only to form a root by which to fix itself to the ground, when it becomes a perfect though diminutive vegetable, capable of providing for itself. This root can be formed only out of proper material: neither water nor anything else which the plantlet is imbibing from the earth will answer the purpose. The proper material is nourishing matter, or prepared food, more or less of which is always provided by the parent plant, and stored up in the seed, either *in* the embryo itself, or *around* it. In the Maple, this nourishment is stored up in the thickish cotyledons, or seed-leaves. And there is barely enough of it to make the beginning of a root, and to provide for the lengthening of the stemlet so as to bring up the unfolding seed-leaves where they may expand to the light of day. But when this is done,

S & F—2

the tiny plant is already able to shift for itself; — that is, to live and continue its growth on what it now takes from the soil and from the air, and *elaborates into nourishment* in its two green leaves, under the influence of the light of the sun.

25. In most ordinary plants, a larger portion of nourishment is provided beforehand in the seed; and the plantlet consequently is not so early or so entirely left to its own resources. Let us examine a number of cases, selected from very common plants. Sometimes, as has just been stated, we find this

26. **Deposit of Food in the Embryo itself.** And we may observe it in every gradation as to quantity, from the Maple of our first illus-

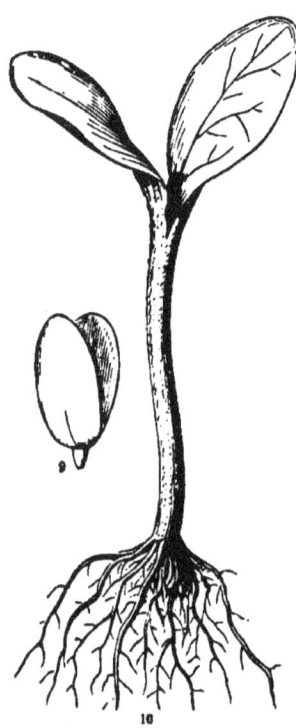

tration, where there is very little, up to the Pea and the Horsechestnut, where there is as much as there possibly can be. If we strip off the coats from the large and flat seed of a Squash or Pumpkin, we find nothing but the embryo within (Fig. 9); and almost the whole bulk of this consists of the two seed-leaves. That these contain a good supply of nourishing matter, is evident from their sweet taste and from their thickness, although there is not enough to obscure their leaf-like appearance. It is by feeding on this supply of nourishment that the germinating Squash or Pumpkin (Fig. 10) grows so rapidly and so vigorously from the seed, — lengthening its stemlet to more than twenty times the length it had in the seed, and thickening it in proportion, — sending out at once a number of roots from its lower end, and soon developing the plumule (16) from its upper end into a third leaf: meanwhile the two cotyledons, relieved from the nourishment with which their tissue was gorged, have expanded into useful green leaves.

27. For a stronger instance, take next the seed of a Plum or Peach, or an Almond, or an Apple-seed (Fig. 11, 12), which shows

FIG. 9. Embryo of a Pumpkin, of the natural size; the cotyledons a little opened
10. The same, when it has germinated.

the same thing on a smaller scale. The embryo, which here also

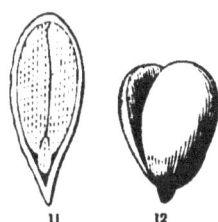

makes up the whole bulk of the kernel of the seed, differs from that of the Pumpkin only in having the seed-leaves more thickened, by the much larger quantity of nourishment stored up in their tissue, — so large and so pure indeed, that the almond becomes an article of food. Fed by this abundant supply, the second, and even the third joints of the stem, with their leaves, shoot forth as soon as the stemlet comes to the surface of the soil. The Beech-nut (Fig. 13), with its sweet and eatable kernel, consisting mainly of a pair of seed-leaves folded together, and gorged with nourishing matter, offers another instance of the same sort : this ample store to feed upon enables the germinating plantlet to grow with remarkable vigor, and to develop a second joint of stem, with its pair of leaves (Fig. 14), before the first pair has expanded or the root has obtained much foothold in the soil.

28. A Bean affords a similar and more familiar illustration. Here the cotyledons in the seed (Fig. 16) are so thick, that, although they are raised out of ground in the ordinary way in germination (Fig. 17), and turn greenish, yet they never succeed in becoming leaf-like, — never display their real nature of leaves, as they do so plainly in the Maple (Fig. 5), the Pumpkin (Fig. 10), the Morning-Glory (Fig. 8, 26 – 28), &c. Turned to great account as magazines of food for the germinating plantlet, they fulfil this special office admirably, but

FIG. 11. An Apple-seed cut through lengthwise, showing the embryo with its thickened cotyledons. 12. The embryo of the Apple, taken out whole, its cotyledons partly separated. FIG. 13. A Beech-nut, cut across. 14. Beginning germination of the Beech, showing the plumule growing before the cotyledons have opened or the root has scarcely formed. 15. The same, a little later, with the second joint lengthened.

they were so gorged and, as it were, misshapen, that they became quite unfitted to perform the office of foliage. This office is accordingly first performed by the succeeding pair of leaves, those of the plumule (Fig. 17, 18), which is put into rapid growth by the abundant nourishment contained in the large and thick seed-leaves. The latter, having fulfilled this office, soon wither and fall away.

29. This is carried a step farther in the Pea (Fig. 19, 20), a near relative of the Bean, and in the Oak (Fig. 21, 22), a near relative of the Beech. The difference in these and many other similar cases is this.

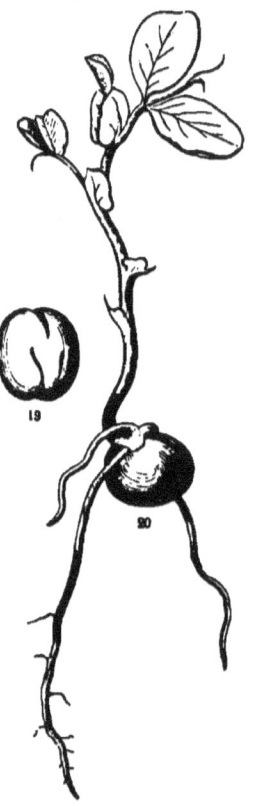

The cotyledons, which make up nearly the whole bulk of the seed are excessively thickened, so as to become nearly hemispherical in shape. They have lost all likeness to leaves, and all power of ever fulfilling the office of leaves. Accordingly in germination they remain unchanged within the husk or coats of the seed, never growing themselves, but supplying abundant nourishment to the plumule (the bud for the forming stem) between them. This pushes forth from the seed, shoots upward, and gives rise

FIG. 16. A Bean: the embryo, from which seed-coats have been removed: the small stem is seen above, bent down upon the edge of the thick cotyledons. 17. The same in early germination; the plumule growing from between the two seed-leaves. 18. The germination more advanced, the two leaves of the plumule unfolded, and raised on a short joint of stem.

FIG. 19. A Pea: the embryo, with the seed-coats taken off. 20. A Pea in germination.

to the first leaves that appear.   In most cases of the sort, the radicle,
or short original stemlet of the embryo be-
low the cotyledons (which is plainly shown
in the Pea, Fig. 19), lengthens very little,
or not at all; and so the cotyledons remain
under ground, if the seed was covered by
the soil, as every one knows to be the case
with Peas.   In these (Fig. 20), as also in
the Oak (Fig. 22), the leaves of the first
one or two joints are imperfect, and mere
small scales; but genuine leaves immediately follow.   The Horsechestnut and Buckeye (Fig. 23, 24) furnish another instance
of the same sort.   These trees are nearly
related to the Maple; but while the seed-
leaves of the Maple show themselves to
be leaves, even in the seed (as we have
already seen), and when they germinate
fulfil the office of ordinary leaves, those
of the Buckeye and of the Horsechestnut
(Fig. 23), would never be suspected to be
the same organs.   Yet they are so, only
in another shape, — exceedingly thickened
by the accumulation of a great quantity
of starch and other nourishing matter in
their substance; and besides, their contiguous faces stick together more or less firmly,
so that they never open.   But the stalks
of these seed-leaves grow, and, as they
lengthen, push the radicle and the plumule
out of the seed, when the former develops downwardly the root, the
latter upwardly the leafy stem and all it bears (Fig. 24).

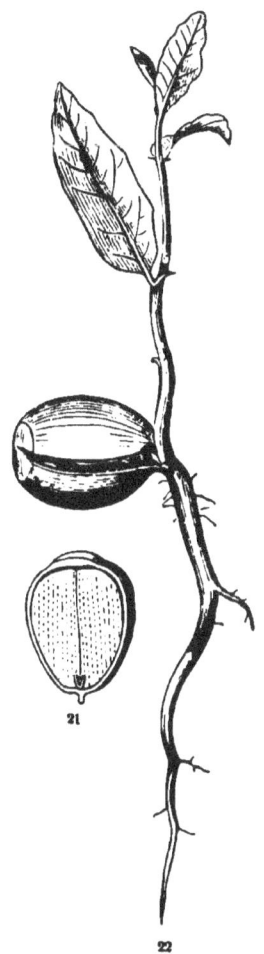

21

22

30. **Deposit of Food outside of the Embryo.**   Very often the nourishment provided for the seedling plantlet is laid up, not *in* the embryo
itself, but *around* it.   A good instance to begin with is furnished by
the common Morning-Glory, or Convolvulus.   The embryo, taken
out of the seed and straightened, is shown in Fig. 26.   It consists
of a short stemlet and of a pair of very thin and delicate green
leaves, having no stock of nourishment in them for sustaining the

FIG. 21.   An acorn divided lengthwise.   22.   The germinating Oak.

2

earliest growth.   On cutting open the seed, however, we find this embryo (considerably crumpled or folded together, so as to occupy

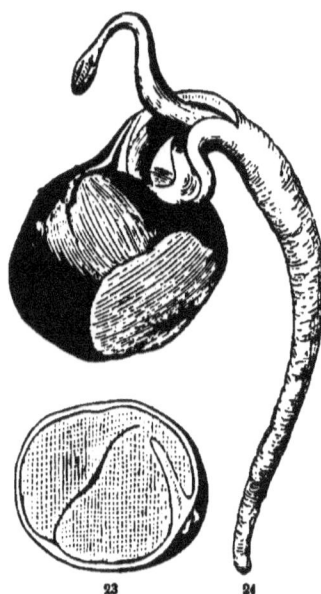

less space, Fig. 25) to be surround-ed by a mass of rich, mucilaginous matter (becoming rather hard and solid when dry), which forms the principal bulk of the seed.   Upon this stock the embryo feeds in ger-mination ; the seed-leaves absorbing it into their tissue as it is rendered soluble (through certain chemical changes) and dissolved by the wa-ter which the germinating seed im-bibes from the moist soil.   Having by this aid lengthened its radicle into a stem of consider-able length,

and formed the beginning of a root at its lower end, already imbedded in the soil (Fig. 27), the cotyledons now disengage themselves from the seed-coats, and ex-pand in the light as the first pair of leaves (Fig. 28).   These immediately begin to elaborate, under the sun's influence, what the root imbibes from the soil, and the new nourishment so produced is used, partly to increase the size of the little stem, root, and leaves already existing, and partly to produce a second joint of stem with its leaf (Fig. 29), then a third with its leaf (Fig. 8) ; and so on.

31. This maternal store of food, deposited in the seed along with the embryo (but not in its substance), the old botanists likened to

the *albumen*, or white of the egg, which encloses the yolk, and therefore gave it the same name,— the *albumen* of the seed,— a name which it still retains. Food of this sort for the plant is also food for animals, or for man ; and it is this albumen, the floury part of the seed, which forms the principal bulk of such important grains as those of Indian Corn (Fig. 38 – 40), Wheat, Rice, Buck-wheat, and of the seed of Four-o'clock, (Fig. 36, 37), and the like. In all these last-named cases, it may be ob-served that the embryo is not enclosed in the albumen, but placed on one side of it, yet in close contact with it, so that the embryo may absorb readily from it the nourishment it requires when it begins to grow. Sometimes

the embryo is coiled around the outside, in the form of a ring, as in the Purslane and the Four-o'clock (Fig. 36, 37) ; sometimes it is coiled within the albumen, as in the Potato (Fig. 34, 35) ; some-times it is straight in the centre of the albumen, occupying nearly its

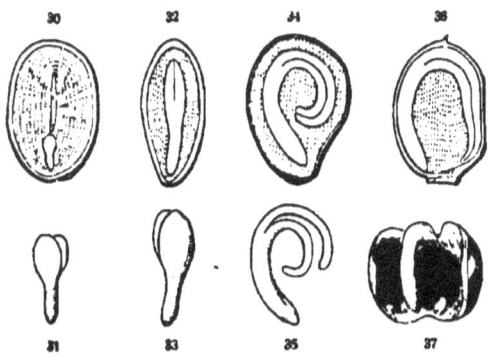

whole length, as in the Barberry (Fig. 32, 33), or much smaller and near one end, as in the Iris (Fig. 43) ; or some-times so minute, in the midst of the al-bumen, that it needs a magnifying-glass to find it, as in the But-

FIG. 29. Germination of the Morning Glory more advanced : the upper part only , showing the leafy cutyledons, the second joint of stem with its leaf, and the third with its leaf just developing.

FIG. 30. Section of a seed of a Peony, showing a very small embryo in the albumen, near one end. 31. This embryo detached, and more magnified.

FIG. 32. Section of a seed of Barberry, showing the straight embryo in the middle of the albumen. 33. Its embryo detached.

FIG. 34. Section of a Potato-seed, showing the embryo coiled in the albumen. 35. Its embryo detached.

FIG. 36. Section of the seed of Four-o'clock, showing the embryo coiled round the outside of the albumen. 37. Its embryo detached.

tercup or the Columbine, and in the Peony (Fig. 30, 31), where, however, it is large enough to be distinguished by the naked eye. Nothing is more curious than the various shapes and positions of the embryo in the seed, nor more interesting than to watch its development in germination. One point is still to be noticed, since the botanist considers it of much importance, namely : —

32. **The Kinds of Embryo as to the Number of Cotyledons.** In all the figures, it is easy to see that the embryo, however various in shape, is constructed on one and the same plan ; — it consists of a radicle or stemlet, with a pair of cotyledons on its summit. Botanists therefore call it *dicotyledonous*, — an inconveniently long word to express the fact that the embryo has two cotyledons or seed-leaves. In many cases (as in the Buttercup), the cotyledons are indeed so minute, that they are discerned only by the nick in the upper end of the little embryo; yet in germination they grow into a pair of seed-leaves, just as in other cases where they are plain to be seen, as leaves, in the seed. But in Indian Corn (Fig. 40), in Wheat, the Onion, the Iris (Fig. 43), &c., it is well known that only one

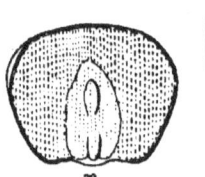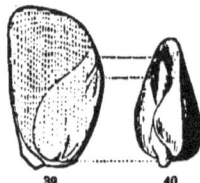

leaf appears at first from the sprouting seed: in these the embryo has only one cotyledon, and it is therefore termed by the botanists *monocotyledonous ;* — an extremely long word, like the other, of Greek derivation, which means *one-cotyledoned.* The rudiments of one or more other leaves are, indeed, commonly present in this sort of embryo, as is plain to see in Indian Corn (Fig. 38 – 40), but they form a bud situated above or within the cotyledon, and enclosed by it more or less completely ; so that they evidently belong to the plumule (16) ; and these leaves appear in the seedling plantlet, each from within its predecessor, and therefore originating higher up on the forming stem (Fig. 42, 44). This will readily be understood from the accompanying figures, with their explanation, which the student may without difficulty verify for him-

FIG. 38. A grain of Indian Corn, flatwise, cut away a little, so as to show the embryo, lying on the albumen, which makes the principal bulk of the seed.

FIG. 39. Another grain of Corn, cut through the middle in the opposite direction, dividing the embryo through its thick cotyledon and its plumule, the latter consisting of two leaves, one enclosing the other.

FIG. 40. The embryo of Corn, taken out whole : the thick mass is the cotyledon ; the narrow body partly enclosed by it is the plumule ; the little projection at its base is the very short radicle enclosed in the sheathing base of the first leaf of the plumule.

self, and should do so, by examining grains of Indian Corn, soaked in water, before and also during germination. In the Onion, Lily, and the Iris (Fig. 43), the monocotyledonous embryo is simpler, consisting apparently of a simple oblong or cylindrical body, in which no distinction of parts is visible : the lower end is *radicle*, and from it grows the root; the rest is a *cotyledon*, which has wrapped up in it a minute *plumule*, or bud, that shows itself when the seeds sprout in germination. The first leaf which appears above ground in all these cases is not the cotyledon. In all seeds with one cotyledon to the embryo, this remains in the seed, or at least its upper part, while its lengthening base comes out, so as to extricate the plumule, which shoots upward, and develops the first leaves of the plantlet. These appear one 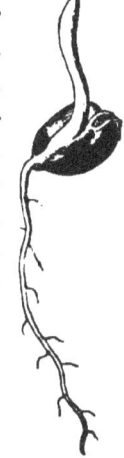 above or within the other in succession, — as is shown in Fig. 42 and Fig. 44, — the first commonly in the form of a little scale or imperfect leaf; the second or third and the following ones as the real, ordinary leaves of the plant. Meanwhile, from the root end of the embryo, a root (Fig. 41, 44), or soon a whole cluster of roots (Fig. 42), makes its appearance.

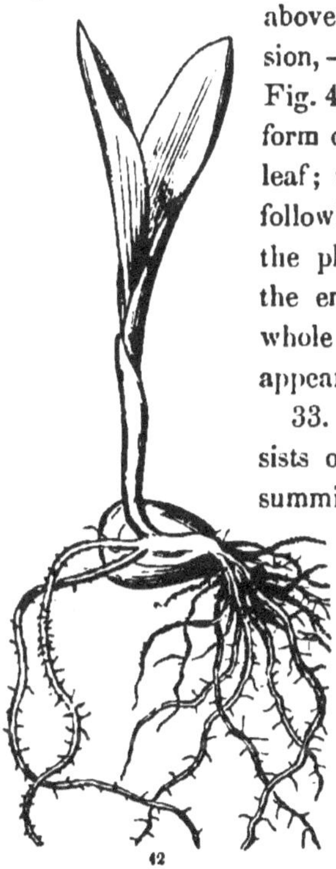

33. In Pines, and the like, the embryo consists of a radicle or stemlet, bearing on its summit three or four, or often from five to ten slender cotyledons, arranged in a cirele (Fig. 45), and expanding at once into á circle of as many green leaves in germination (Fig. 46). Such embryos are said to be *polycotyledonous*, that is, as the word denotes, many-cotyledoned.

34. **Plan of Vegetation.** The student who has understandingly followed the growth of the embryo in the seed into the seedling plantlet, — composed of a root, and a stem of two or three joints, each bearing a

FIG. 41. Grain of Indian Corn in germination.
FIG. 42. The same, further advanced.

leaf, or a pair (rarely a circle) of leaves, — will have gained a correct idea of the plan of vegetation in general, and have laid a good foundation for a knowledge of the whole structure and physiology of plants. For the plant goes on to grow in the same way throughout, by mere repetitions of what the early germinating plantlet displays to view, — of what was contained, in miniature or in rudiment, in the seed itself. So far as vegetation is concerned (leaving out of view for the present the flower and fruit), the full-grown leafy herb or tree, of whatever size, has nothing, and does nothing, which the seedling plantlet does not have and do. The whole mass of stem or trunk and foliage of the complete plant, even of the largest forest-tree, is composed of a succession or multiplication of similar parts, — one arising from the summit of another, — each, so to say, the offspring of the preceding and the parent of the next.

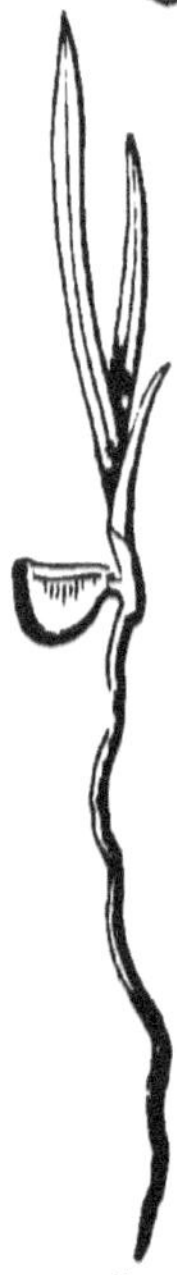

35. In the same way that the earliest portions of the seedling stem, with the leaves they bear, are successively produced, so, joint by joint in direct succession, a single, simple, leafy stem is developed and carried up. Of such a simple leafy stem many a plant consists (before flowering, at least), — many herbs, such as Sugar-Cane, Indian Corn, the Lily, the tall Banana, the Yucca, &c.; and among trees the Palms and the Cycas (wrongly called Sago Palm) exhibit the same simplicity, their stems, of whatever age, being unbranched columns (Fig. 47). (Growth in diameter is of course to be considered, as well as growth in length. That, and the question *how* growth of any kind takes place, we will consider hereafter.) But more commonly, as soon as the plant has produced a main stem of a certain length, and displayed a certain amount of foliage, it begins to

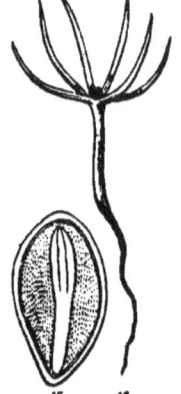

FIG. 43. Section of a seed of the Iris, or Flower-de-Luce, showing its small embryo in the albumen, near the bottom.

FIG. 44. Germinating plantlet of the Iris.

FIG. 45. Section of a seed of a Pine, with its embryo of several cotyledons. 46. Early seedling Pine, with its stemlet, displaying its six seed-leaves.

produce additional stems, that is, *branches.*   The branching plant we will consider in the next Lesson.

36. The subjoined figures (Fig. 47) give a view of some forms of *simple-stemmed* vegetation.   The figure in the foreground on the left represents a Cycas (wrongly called in the conservatories Sago Palm).   Behind it is a Yucca (called Spanish Bayonet at the South) and two Cocoanut Palm-trees.   On the right is some Indian Corn, and behind it a Banana.

47

LESSON IV.

THE GROWTH OF PLANTS FROM BUDS AND BRANCHES.

37. WE have seen how the plant grows so as to produce a root, and a simple stem with its foliage. Both the root and stem, however, generally branch.

38. The branches of the root arise without any particular order. There is no telling beforehand from what part of a main root they will spring. But the branches of the stem, except in some extraordinary cases, regularly arise from a particular place. Branches or shoots in their undeveloped state are

39. **Buds.** These regularly appear in the *axils* of the leaves, — that is, in the angle formed by the leaf with the stem on the upper side ; and as leaves are symmetrically arranged on the stem, the buds, and the branches into which the buds grow, necessarily partake of this symmetry.

40. We do not confine the name of bud to the scaly winter-buds which are so conspicuous on most of our shrubs and trees in winter and spring. It belongs as well to the forming branch of any herb, at its first appearance in the axil of a leaf. In growing, buds lengthen into branches, just as the original stem did from the plumule of the embryo (16) when the seed germinated. Only, while the original stem is implanted in the ground by its root, the branch is implanted on the stem. Branches, therefore, are repetitions of the main stem. They consist of the same parts, — namely, joints of stem and leaves, — growing in the same way   And in the axils of their leaves another crop of buds is naturally produced, giving rise to another generation of branches, which may in turn produce still another generation ; and so on, — until the tiny and simple seedling develops into a tall and spreading herb or shrub ; or into a massive tree, with its hundreds of annually increasing branches, and its thousands, perhaps millions, of leaves.

41. The herb and the tree grow in the same way. The difference is only in size and duration.

An *Herb* dies altogether, or dies down to the ground, after it has ripened its fruit, or at the approach of winter.

An *annual herb* flowers in the first year, and dies, root and all, after ripening its seed : Mustard, Peppergrass, Buckwheat, &c., are examples.

A *biennial herb* — such as the Turnip, Carrot, Beet, and Cabbage — grows the first season without blossoming, survives the winter, flowers after that, and dies, root and all, when it has ripened its seed.

A *perennial herb* lives and blossoms year after year, but dies down to the ground, or near it, annually, — not, however, quite down to the root : for a portion of the stem, with its buds, still survives ; and from these buds the shoots of the following year arise.

A *Shrub* is a perennial plant, with woody stems which continue alive and grow year after year.

A *Tree* differs from a shrub only in its greater size.

42. **The Terminal Bud.** There are herbs, shrubs, and trees which do not branch, as we have already seen (35) ; but whose stems, even when they live for many years, rise as a simple shaft (Fig. 47). These plants grow by the continued evolution of a bud which crowns the summit of the stem, and which is therefore called the *terminal bud.* This bud is very conspicuous in many branching plants also ; as on all the stems or shoots of Maples (Fig. 53), Horsechestnuts (Fig. 48), or Hickories (Fig. 49), of a year old. When they grow, they merely prolong the shoot or stem on which they rest. On these same shoots, however, other buds are to be seen, regularly arranged down their sides. We find them situated just over broad, flattened places, which are the scars left by the fall of the leaf-stalk the autumn previous. Before the fall of the leaf, they would have been seen to occupy their *axils* (39) : so they are named

43. **Axillary Buds.** They were formed in these trees early in the summer. Occasionally they grow at the time into branches : at least, some of them are pretty sure to do so, in case the growing terminal bud at the end of the shoot is injured or destroyed. Otherwise they lie dormant until the spring. In many trees or shrubs (such for example as the Sumach and Honey-Locust) these axillary buds do not show themselves until spring ; but if

FIG. 48.   Shoot of Horsechestnut, of one year's growth, taken in autumn after the leaves have fallen.

searched for, they may be detected, though of small size, hidden under the bark. Sometimes, although early formed, they are concealed all summer long under the base of the leaf-stalk, hollowed out into a sort of inverted cup, like a candle-extinguisher, to cover them; as in the Locust, the Yellow-wood, or more strikingly in the Button-wood or Plane-tree (Fig. 50).

44. Such large and conspicuous buds as those of the Horsechestnut, Hickory, and the like, are *scaly;* the scales being a kind of imperfect leaves. The use of the bud-scales is obvious; namely, to protect the tender young parts beneath. To do this more effectually, they are often coated on the outside with a varnish which is impervious to wet, while within they, or the parts they enclose, are thickly clothed with down or wool; not really to keep out the cold of winter, which will of course penetrate the bud in time, but to shield the interior against sudden changes from warm to cold, or from cold to warm, which are equally injurious. Scaly buds commonly belong, as would be expected, to trees and shrubs of northern climates; while *naked* buds are usual in tropical regions, as well as in herbs everywhere which branch during the summer's growth and do not endure the winter.

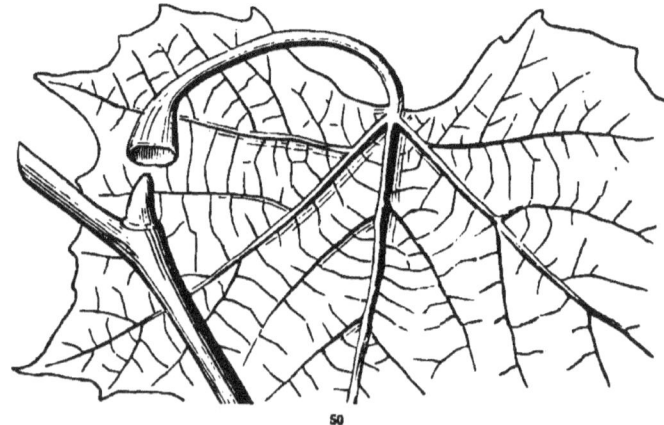

50

45. But *naked buds*, or nearly naked, also occur in several of oui own trees and shrubs; sometimes pretty large ones, as those of Hob

FIG. 49. Annual shoot of the Shagbark Hickory.
FIG. 50. Bud and leaf of the Buttonwood, or American Plane-tree.

blebush (while those of the nearly-related Snowball or High Bush-Cranberry are scaly); but more commonly, when naked buds occur in trees and shrubs of our climate, they are small, and sunk in the bark, as in the Sumac; or even partly buried in the wood until they begin to grow, as in the Honey-Locust.

46. **Vigor of Vegetation from Buds.** Large and strong buds, like those of the Horsechestnut, Hickory, and the like, on inspection will be found to contain several leaves, or pairs of leaves, ready formed, folded and packed away in small compass, just as the seed-leaves are packed away in the seed: they even contain all the blossoms of the ensuing season, plainly visible as small buds. And the stems upon which these buds rest are filled with abundant nourishment, which was deposited the summer before in the wood or in the bark. Under the surface of the soil, or on it, covered with the fallen leaves of autumn, we may find similar strong buds of our perennial herbs, in great variety; while beneath are thick roots, rootstocks, or tubers, charged with a great store of nourishment for their use. As we regard these, we shall readily perceive how it is that vegetation shoots forth so vigorously in the spring of the year, and clothes the bare and lately frozen surface of the soil, as well as the naked boughs of trees, almost at once with a covering of the freshest green, and often with brilliant blossoms. Everything was prepared, and even formed, beforehand: the short joints of stem in the bud have only to lengthen, and to separate the leaves from each other so that they may unfold and grow. Only a small part of the vegetation of the season comes directly from the seed, and none of the earliest vernal vegetation. This is all from buds which have lived through the winter.

47. This growth from buds, in manifold variety, is as interesting a subject of study as the growth of the plantlet from the seed, and is still easier to observe. We have only room here to sketch the general plan; earnestly recommending the student to examine attentively their mode of growth in all the common trees and shrubs, when they shoot forth in spring. The growth of the terminal bud prolongs the stem or branch: the growth of axillary buds produces branches.

48. **The Arrangement of Branches** is accordingly the same as of axillary buds; and the arrangement of these buds is the same as that of the leaves. Now leaves are arranged in two principal ways: they are either *opposite* or *alternate*. Leaves are *opposite* when

there are two borne on the same joint of stem, as in the Horse-chestnut, Maple (Fig. 7), Honeysuckle (Fig. 132), Lilac, &c.; the two leaves in such cases being always *opposite* each other, that is, on exactly opposite sides of the stem. Here of course the buds in their axils are opposite, as we observe in Fig. 48, where the leaves have fallen, but their place is shown by the scars. And the branches into which the buds grow are likewise opposite each other in pairs.

49. Leaves are *alternate* when there is only one from each joint of stem, as in the Oak (Fig. 22), Lime-tree, Poplar, Buttonwood (Fig. 50), Morning-Glory (Fig. 8), — not counting the seed-leaves, which of course are opposite, there being a pair of them; also in Indian Corn (Fig. 42), and Iris (Fig. 44). Consequently the axillary buds are also alternate, as in Hickory (Fig. 49); and the branches they form alternate, — making a different kind of spray from the other mode, — one branch shooting on the one side of the stem and the next on some other. For in the alternate arrangement no leaf is on the same side of the stem as the one next above or next below it.

50. Branches, therefore, are arranged with symmetry; and the mode of branching of the whole tree may be foretold by a glance at the arrangement of the leaves on the seedling or stem of the first year. This arrangement of the branches according to that of the leaves is always plainly to be recognized; but the symmetry of branches is rarely complete. This is owing to several causes; mainly to one, viz.: —

51. It never happens that all the buds grow. If they did, there would be as many branches in any year as there were leaves the year before. And of those which do begin to grow, a large portion perish, sooner or later, for want of nourishment or for want of light. Those which first begin to grow have an advantage, which they are apt to keep, taking to themselves the nourishment of the stem, and starving the weaker buds.

52. In the Horsechestnut (Fig. 48), Hickory (Fig. 49), Mag-nolia, and most other trees with large scaly buds, the terminal bud is the strongest, and has the advantage in growth, and next in strength are the upper axillary buds: while the former continues the shoot of the last year, some of the latter give rise to branches, while the rest fail to grow. In the Lilac also, the upper axillary buds are stronger than the lower; but the terminal bud rarely

appears at all; in its place the uppermost pair of axillary buds grow, and so each stem branches every year into two; making a repeatedly two-forked ramification.

53. In these and many similar trees and shrubs, most of the shoots make a *definite annual growth*. That is, each shoot of the season develops rapidly from a strong bud in spring, — a bud which generally contains, already formed in miniature, all or a great part of the leaves and joints of stem it is to produce, — makes its whole growth in length in the course of a few weeks, or sometimes even in a few days, and then forms and ripens its buds for the next year's similar rapid growth.

54. On the other hand, the Locust, Honey-Locust, Sumac, and, among smaller plants, the Rose and Raspberry, make an *indefinite annual growth*. That is, their stems grow on all summer long, until stopped by the frosts of autumn or some other cause; consequently they form and ripen no terminal bud protected by scales, and the upper axillary buds are produced so late in the season that they have no time to mature, nor has the wood time to solidify and ripen. Such stems therefore commonly die at the top in winter, or at least all their upper buds are small and feeble; and the growth of the succeeding year takes place mainly from the lower axillary buds, which are more mature. Most of our perennial herbs grow in this way, their stems dying down to the ground every year: the part beneath, however, is charged with vigorous buds, well protected by the kindly covering of earth, ready for the next year's vegetation.

55. In these last-mentioned cases there is, of course, no single main stem, continued year after year in a direct line, but the trunk is soon lost in the branches; and when they grow into trees, these commonly have rounded or spreading tops. Of such trees with *deliquescent* stems, — that is, with the trunk dissolved, as it were, into the successively divided branches, the common American Elm (Fig. 54) furnishes a good illustration.

56. On the other hand, the main stem of Pines and Spruces, as it begins in the seedling, unless destroyed by some injury, is carried on in a direct line throughout the whole growth of the tree, by the development year after year of a terminal bud: this forms a single, uninterrupted shaft, — an *excurrent* trunk, which can never be confounded with the branches that proceed from it. Of such *spiry* or *spire-shaped* trees, the Firs or Spruces are the most perfect and

3

familiar illustrations (Fig. 54); but some other trees with strong terminal buds exhibit the same character for a certain time, and in a less marked degree.

57. **Latent Buds.** Some of the axillary buds grow the following year into branches; but a larger number do not (51). These do not necessarily die. Often they survive in a latent state for some years, visible on the surface of the branch, or are smaller and concealed under the bark, resting on the surface of the wood: and when at any time the other buds or branches happen to be killed, these older latent buds grow to supply their place; — as is often seen when the foliage and young shoots of a tree are destroyed by insects. The new shoots seen springing directly out of large stems may sometimes originate from such latent buds, which have preserved their life for years. But commonly these arise from

58. **Adventitious Buds.** These are buds which certain shrubs and trees produce anywhere on the surface of the wood, especially where it has been injured. They give rise to the slender twigs which often feather so beautifully the sides of great branches or trunks of our American Elms. They sometimes form on the root, which naturally is destitute of buds; and they are sure to appear on the trunks and roots of Willows, Poplars, and Chestnuts, when these are wounded or mutilated. Indeed Osier-Willows are *pollarded*, or cut off, from time to time, by the cultivator, for the purpose of producing a crop of slender adventitious twigs, suitable for basket-work. Such branches, being altogether irregular, of course interfere with the natural symmetry of the tree (50). Another cause of irregularity, in certain trees and shrubs, is the formation of what are called

59. **Accessory or Supernumerary Buds.** There are cases where two,
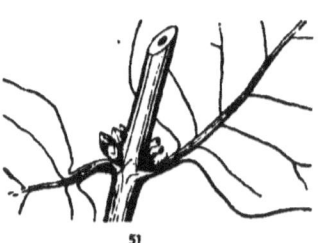
three, or more buds spring from the axil of a leaf, instead of the single one which is ordinarily found there. Sometimes they are placed one over the other, as in the Aristolochia or Pipe-Vine, and in the Tartarian Honeysuckle (Fig. 51); also in the Honey-Locust, and in the Walnut and Butternut (Fig. 52), where the upper supernumerary bud is a good way out of the axil and above the others. And this is here stronger

FIG. 51. Tartarian Honeysuckle, with three accessory buds in one axil.

than the others, and grows into a branch which is considerably out of the axil, while the lower and smaller ones commonly do not grow at all.   In other cases the three buds stand side by side in the axil, as in the Hawthorn, and the Red Maple (Fig. 53).   If these were all to grow into branches, they would stifle or jostle each other.   But some of them are commonly flower-buds : in the Red Maple, only the middle one is a leaf-bud, and it does not grow until after those on each side of it have expanded the blossoms they contain.

60. **Sorts of Buds.** It may be useful to enumerate the kinds of buds which have now been mentioned, referring back to the paragraphs in which the peculiarities of each are explained.   Buds, then, are either *terminal* or *lateral.* They are

Terminal when they rest on the apex of a stem (42).   The earliest terminal bud is the *plumule* of the embryo (16).

Lateral, when they appear on the side of a stem : — of which the only regular kind is the

Axillary (43), namely, those which are situated in the axils of leaves.

Accessory or *Supernumerary* (59), when two or more occur in addition to the ordinary axillary bud.

Adventitious (58), when they occur out of the axils and without order, on stems or roots, or even on leaves.   Any of these kinds may be, either

Naked, when without coverings; or *scaly,* when protected by scales (44, 45).

Latent, when they survive long without growing, and commonly without being visible externally (57).

Leaf-buds, when they contain leaves, and develop into a leafy shoot.

Flower-buds, when they contain blossoms, and no leaves, as the

FIG. 52.   Butternut branch, with accessory buds, the uppermost above the axil.
FIG. 53.   Red-Maple branch, with accessory buds placed side by side.

side-buds of the Red-Maple, or when they are undeveloped blossoms. These we shall have to consider hereafter.

Figure 54 represents a spreading-topped tree (American Elm), the stem dividing off into branches; and some spiry trees (Spruces on the right hand, and two of the Arbor-Vitæ on the left) with excurrent stems.

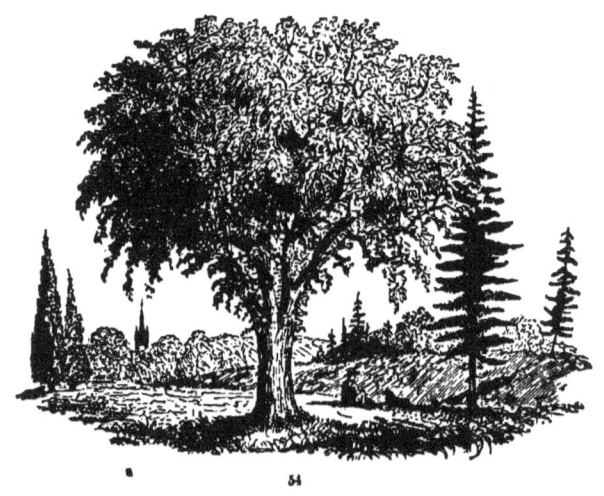

54

LESSON V.

MORPHOLOGY (i.e. VARIOUS SORTS AND FORMS) OF ROOTS.

61. **Morphology,** as the name (derived from two Greek words) denotes, is the doctrine of forms. In treating of forms in plants, the botanist is not confined to an enumeration or description of the shapes or sorts that occur, — which would be a dull and tedious business. — but he endeavors to bring to view *the relations between one form and another;* and this is an interesting study.

62. Botanists give particular names to all the parts of plants, and also particular terms to express their principal varieties in form. They use these terms with great precision and advantage in describing the species or kinds of plants. They must therefore be defined and explained in our books. But it would be a great waste of time

for the young student to learn them by rote. The student should
rather consider the connection between one form and another; and
notice how the one simple plan of the plant, as it has already been
illustrated, is worked out in the greatest variety of ways, through the
manifold diversity of forms which each of its three organs of vegetation — root, stem, and leaf — is made to assume.

63. This we are now ready to do. That is, having obtained a
g neral idea of vegetation, by tracing the plant from the seed and
the bud into the herb, shrub, or tree, we proceed to contemplate the
principal forms under which these three organs occur in different
plants, or in different parts of the same plant; or, in other words, to
study the *morphology* of the root, stem, and leaves.

64. Of these three organs, the root is the simplest and the least
varied in its modifications. Still it exhibits some widely different
kinds. Going back to the beginning, we commence with

65. **The simple Primary Root,** which most plants send down from
the root-end of the embryo as it grows from the seed; as we have
seen in the Maple (Fig 5 – 7), Morning-Glory (Fig. 8 and 28),
Beech (Fig. 14, 15), Oak and Buckeye (Fig. 22 – 24), &c. This,
if it goes on to grow, makes a *main* or *tap* root, from which side-
branches here and there proceed. Some plants keep this main root
throughout their whole life, and send off only small side bra' ches;
as in the Carrot (Fig. 58) and Radish (Fig. 59): and in some trees,
like the Oak, it takes the lead of the side-branches for many years,
unless accidentally injured, as a strong tap-root. But commonly
the main root divides off very soon, and is lost in the branches.
We have already seen, also, that there may be at the beginning

66. **Multiple Primary Roots.** We have noticed them in the Pumpkin (Fig. 10), in the Pea (Fig. 20), and in Indian Corn (Fig. 42).
That is, several roots have started all at once, or nearly so, from the
seedling stem, and formed a bundle or cluster (a *fascicled* root, as
it is called), in place of one main root. The Bean, as we observe
in Fig. 18, begins with a main root , but some of its branches soon
overtake it, and a cluster of roots is formed.

67. **Absorption of Moisture by Roots.** The branches of roots as they
grow commonly branch again and again, into smaller roots or *rootlets ;*
in this way very much increasing the surface by which the plant
connects itself with the earth, and absorbs moisture from it. The
whole surface of the root absorbs, so long as it is fresh and new ;
and the newer the roots and rootlets are, the more freely do they

3 *

imbibe.   Accordingly, as long as the plant grows above ground, and
expands fresh foliage, from which moisture much of the time largely
escapes into the air, so long it continues to extend and multiply its
roots in the soil beneath, renewing and increasing the fresh surface
for absorbing moisture, in proportion to the demand from above.
And when growth ceases above ground, and the leaves die and fall,
or no longer act, then the roots generally stop growing, and their
soft and tender tips harden.   From this period, therefore, until
growth begins anew the next spring, is the best time for transplant-
ing; especially for trees and shrubs, and herbs so large that they
cannot well be removed without injuring the roots very mnch.

68. We see, on considering a moment, that an herb or a tree
consists of two great surfaces, with a narrow part or trunk between
them, — one surface spread out in the air, and the other in the soil.
These two surfaces bear a certain proportion to each other; and the

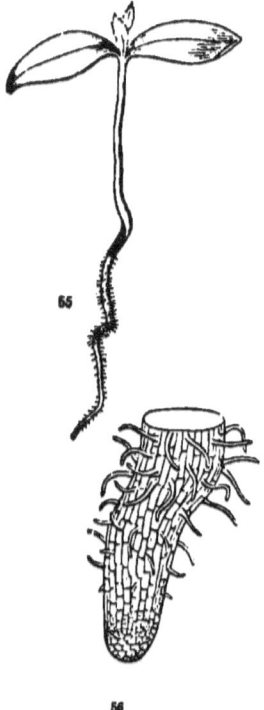

upper draws largely on the lower for
moisture.   Now, when the leaves fall
from the tree in autumn, the vast surface exposed to the air is reduced to a
very small part of what it was before;
and the remainder, being covered with
a firm bark, cannot lose much by evaporation.   In common herbs the whole
surface above ground perishes in autumn; and many of the rootlets die at
the same time, or soon afterwards.
So that the living vegetable is reduced
for the time to the smallest compass,
— to the thousandth or hundred-thou-
sandth part of what it was shortly
before, — and what remains alive rests
in a dormant state, and may now be
transplanted without much danger of
harm.   If any should doubt whether
there is so great a difference between
the summer and the winter size of
plants, let them compare a lily-bulb
with the full-grown Lily, or calculate the surface of foliage which

FIG. 55. Seedling Maple, of the natural size, showing the root-hairs. 56. A bit of the
end of the root magnified.

a tree exposes to the air, as compared with the surface of its twigs.

69. The absorbing surface of roots is very much greater than it appears to be, on account of the root-hairs, or slender fibrils, which abound on the fresh and new parts of roots. These may be seen with an ordinary magnifying-glass, or even by the naked eye in many cases; as in the root of a seedling Maple (Fig. 55), where the surface is thickly clothed with them. They are not rootlets of a smaller sort; but, when more magnified, are seen to be mere elongations of the surface of the root into slender tubes, which through their very delicate walls imbibe moisture from the soil with great avidity. They are commonly much longer than those shown in Fig. 56, which represents only the very tip of a root moderately magnified. Small as they are individually, yet the whole amount of absorbing surface added to the rootlets by the countless numbers of these tiny tubes is very great.

70. Roots intended mainly for absorbing branch freely, and are slender or thread-like. When the root is principally of this character it is said to be *fibrous;* as in Indian Corn (Fig. 42), and other grain, and to some extent in all annual plants (41).

71. **The Root as a Storehouse of Food.** In biennial and many perennial herbs (41), the root answers an additional purpose. In the course of the season it becomes a storehouse of nourishment, and enlarges or thickens as it receives the accumulation. Such roots are said to be *fleshy;* and different names are applied to them according to

FIG. 57 58, 59. Forms of fleshy or thickened roots.

their shapes. We may divide them all into two kinds ; 1st, those consisting of one main root, and 2d, those without any main root.

72. The first are merely different shapes of the *tap-root* ; which is

Conical, when it thickens most at the crown, or where it joins the stem, and tapers regularly downwards to a point, as in the Common Beet, the Parsnip, and Carrot (Fig. 58) :

Turnip-shaped or *napiform,* when greatly thickened above ; but abruptly becoming slender below ; as the Turnip (Fig. 57) : and,

Spindle-shaped, or *fusiform,* when thickest in the middle and tapering to both ends ; as the common Radish (Fig. 59).

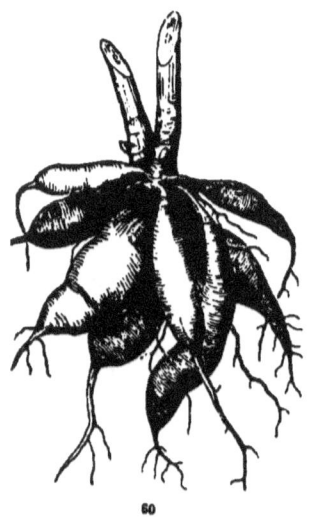

73. In the second kind, where there is no main root, the store of nourishing matter may be distributed throughout the branches or cluster of roots generally, or it may be accumulated in some of them, as we see in the *tuberous* roots of the Sweet Potato, the common Peony, and the Dahlia (Fig. 60).

74. All but the last of these illustrations are taken from *biennial* plants. These grow with a large tuft of leaves next the ground, and accumulate nourishment all the first summer, and store up all they produce beyond what is wanted at the time in their great root, which lives over the winter. We know very well what use man and other animals make of this store of food, in the form of starch, sugar, jelly, and the like. From the second year's growth we may learn what use the plant itself makes of it. The new shoots then feed upon it, and use it to form with great rapidity branches, flower-stalks, blossoms, fruit, and seed ; and, having used it up, the whole plant dies when the seeds have ripened.

75. In the same way the nourishment contained in the separate tuberous roots of the Sweet Potato and the Dahlia (Fig 60) is fed upon in the spring by the buds of the stem they belong to ; and as they are emptied of their contents, they likewise die and decay. But meanwhile similar stores of nourishment, produced by the second year's vegetation, are deposited in new roots, which live through the

FIG. 60. Clustered tuberous roots of the Dahlia, with the bottom of the stem they belong to.

next winter, and sustain the third spring's growth, and so on ; — these plants being *perennial* (41), or lasting year after year, though each particular root lives little more than one year.

76. Many things which commonly pass for roots are not really roots at all. Common potatoes are tuberous parts of stems, while sweet potatoes are roots, like those of the Dahlia (Fig. 60). The difference between them will more plainly appear in the next Lesson.

77. **Secondary Roots.** So far we have considered only the original or primary root, — that which proceeded from the lower end of the first joint of stem in the plantlet springing from the seed, — and its subdivisions. We may now remark, that any other part of the stem will produce roots just as well, whenever favorably situated for it ; that is, when covered by the soil, which provides the darkness and the moisture which is congenial to them. For these *secondary* roots, as they may be called, partake of the ordinary disposition of the organ : they avoid the light, and seek to bury themselves in the ground. In Indian Corn we see roots early striking from the second and the succeeding joints of stem under ground, more abundantly than from the first joint (Fig. 42). And all stems that keep up a connection with the soil — such as those which creep along on or beneath its surface — are sure to strike root from almost every joint. So will most branches when bent to the ground, and covered with the soil : and even cuttings from the branches of most plants car. be made to do so, if properly managed. Propagation by buds depends upon this. That is, a piece of a plant which has stem and leaves, either developed or in the bud, may be made to produce roots, and so become an independent plant.

78. In many plants the disposition to strike root is so strong, that they even will spring from the stem above ground. In Indian Corn, for example, it is well known that roots grow, not only from all those joints round which the earth is heaped in hoeing, but also from those several inches above the soil : and other plants produce them from stems or branches high in the air. Such roots are called

79. **Aerial Roots.** All the most striking examples of these are met with, as we might expect, in warmer and damper climates than ours, and especially in deep forests which shut out much of the light ; this being unfavorable to roots. The Mangrove of tropical shores, which occurs on our own southern borders ; the Sugar Cane, from which roots strike just as in Indian Corn, only from higher up the stem ; the Pandanus, called Screw Pine (not from its resemblance to a

Pine-tree, but because it is like a Pine-apple plant) ; and the famous Banyan of India, and some other Fig-trees, furnish the most remarkable examples of roots, which strike from the stem or the branches in the open air, and at length reach the ground, and bury themselves, when they act in the same manner as ordinary roots.

80. Some of our own common plants, however, produce small *aerial rootlets;* not for absorbing nourishment, but for climbing. By these rootlets, that shoot out abundantly from the side of the stems and branches, the Trumpet Creeper, the Ivy of Europe, and our Poison Rhus, — here called Poison Ivy, — fasten themselves firmly to walls, or the trunks of trees, often ascending to a great height. Here roots serve the same purpose that tendrils do in the Grape-Vine and Virginia Creeper. Another form, and the most aerial of all roots, since they never reach the ground, are those of

81. **Epiphytes, or Air-Plants.** These are called by the first name (which means growing on plants), because they are generally found upon the trunks and branches of trees ; — not that they draw any nourishment from them, for their roots merely adhere to the bark, and they flourish just as well upon dead wood or any other convenient support. They are called *air-plants* because they really live altogether upon what they get from the air, as they have no connection with the soil. Hundreds of air-plants grow all around us without attracting any attention, because they are small or humble. Such are the Lichens and Mosses that abound on the trunks or boughs of trees, especially on the shaded side, and on old walls, fences, or rocks, from which they obtain no nourishment. But this name is commonly applied only to the larger, flower-bearing plants which live in this way. These belong to warm and damp parts of the world, where there is always plenty of moisture in the air. The greater part belong to the Orchis family and to the Pine-Apple family ; and among them are some of the handsomest flowers known. We have two or three flowering air-plants in the Southern States, though they are not showy ones. One of them is an Epidendrum growing on the boughs of the Great-flowered Magnolia: another is the Long-Moss, or Black Moss, so called, — although it is no Moss at all, — which hangs from the branches of Oaks and Pines in all the warm parts of the Southern States. (Fig 61 represents both of these. The upper is the Epidendrum conopseum ; the lower, the Black Moss, Tillandsia usneoides.)

82. **Parasitic Plants** exhibit roots under yet another remarkable

aspect.   For these are not merely fixed upon other plants, as air-plants are, but strike their roots, or what answer to roots, into them, and feed on their juices.   Not only Moulds and Blights (which are plants of very low organization) live in this predacious way, but many flowering herbs, and even shrubs.   One of the latter is the Mistletoe, the seed of which germinates on the bough of the tree where it falls or is left by birds; and the forming root penetrates the bark and engrafts itself into the wood, to which it becomes united as firmly as a natural branch to its parent stem ; and indeed the parasite lives just as if it were a branch of the tree it grows and feeds on. A most common parasitic herb is the Dodder; which abounds in low grounds everywhere in summer, and coils its long and slender leafless, yellowish stems — resembling tangled threads of yarn — round and round the stalks of other plants ; wherever they touch piercing the bark with minute and very short rootlets in the form of suckers, which draw out the nourishing juices of the plants laid hold of.   Other parasitic plants, like the Beech-drops and Pine-sap, fasten their roots under ground upon the roots of neighboring plants, and rob them of their rich juices.

LESSON VI.

MORPHOLOGY OF STEMS AND BRANCHES.

83. The growth of the stem in length, and the formation of branches, have been considered already. Their growth in thickness we may study to more advantage in a later Lesson. The very various forms which they assume will now occupy our attention, — beginning with

84. **The Forms of Stems and Branches above ground.** The principal differences as regards size and duration have been mentioned before (41); namely, the obvious distinction of plants into herbs, shrubs, and trees, which depends upon the duration and size of the stem. The stem is accordingly

Herbaceous, when it dies down to the ground every year, or after blossoming.

Suffrutescent, when the bottom of the stem above the soil is a little woody, and inclined to live from year to year.

Suffruticose, when low stems are decidedly woody below, but herbaceous above.

Fruticose, or *shrubby*, when woody, living from year to year, and of considerable size, — not, however, more than three or four times the height of a man.

Arborescent, when tree-like in appearance, or approaching a tree in size.

Arboreous, when forming a proper tree trunk.

85. When the stem or branches rise above ground and are apparent to view, the plant is said to be *caulescent* (that is, to have a *caulis* or true stem). When there is no evident stem above ground, but only leaves or leaf-stalks and flower-stalks, the plant is said to be *acaulescent*, i. e. *stemless*, as in the Crocus, Bloodroot, common Violets, &c., and in the Beet, Carrot, and Radish (Fig. 59), for the first season. There is a stem, however, in all such cases, only it remains on or beneath the ground, and is sometimes very short. Of course leaves and flowers do not arise from the root. These concealed sorts of stem we will presently study.

86. The direction taken by stems, &c., or their mode of growth,

gives rise to several terms, which may be briefly mentioned: —
such as

Diffuse, when loosely spreading in all directions.

Declined, when turned or bending over to one side.

Decumbent, reclining on the ground, as if too weak to stand.

Assurgent or *ascending,* when rising obliquely upwards.

Procumbent or *prostrate,* lying flat on the ground from the first.

Creeping, or *repent,* when prostrate stems on or just beneath the
ground strike root as they grow ; as does the White Clover, the
little Partridge-berry, &c.

Climbing, or *scandent,* when stems rise by clinging to other objects for support, — whether by *tendrils,* as do the Pea, Grape-
Vine, and Virginia Creeper (Fig. 62) ; by their twisting leaf-stalks,
as the Virgin's Bower ; or by rootlets, like the Ivy, Poison Ivy, and
Trumpet Creeper (80).

Twining, or *voluble,* when stems rise by coiling themselves spirally
around other stems or supports ; like the Morning-Glory and the Bean.

87. Certain forms of stems have received distinct names. The
jointed stem of Grasses and Sedges is called by botanists a *culm ;*
and the peculiar scaly trunk of Palms and the like (Fig. 47) is
sometimes called a *caudex.* A few forms of branches the gardener
distinguishes by particular names ; and they are interesting from
their serving for the natural propagation of plants from buds, and
for suggesting ways by which we artificially multiply plants that
would not propagate themselves without the gardener's aid. These
are *suckers, offsets, stolons,* and *runners.*

88. **Suckers** are ascending branches rising from stems under ground,
such as are produced so abundantly by the Rose, Raspberry, and
other plants said to multiply " by the root." If we uncover them,
we see at once the great difference between these subterranean
branches and real roots. They are only creeping branches under
ground. Remarking how the upright shoots from these branches
become separate plants, simply by the dying off of the connecting
under-ground stems, the gardener expedites the result by cutting
them through with his spade. That is, he propagates the plant " by
division."

89. **Stolons** are trailing or reclining branches above ground, which
strike root where they touch the soil, and then send up a vigorous
shoot, which has roots of its own, and becomes an independent plant
when the connecting part dies, as it does after a while. The Currant

4

and the Gooseberry naturally multiply in this way, as well as by suckers (which we see are just the same thing, only the connecting part is concealed under ground). They must have suggested the operation of *layering*, or bending down and covering with earth branches which do not naturally make stolons ; and after they have taken root, as they almost always will, the gardener cuts through the connecting stem, and so converts a rooting branch into a separate plant.

90. **Offsets**, like those of the Houseleek, are only short stolons, with a crown of leaves at the end.

91. **Runners**, of which the Strawberry presents the most familiar example, are a long and slender, tendril-like, leafless form of creeping branches. Each runner, after having grown to its full length, strikes root from the tip, and fixes it to the ground, then forms a bud there, which develops into a tuft of leaves, and so gives rise to a new plant, which sends out new runners to act in the same way. In this manner a single Strawberry plant will spread over a large space, or produce a great number of plants, in the course of the summer ; — all connected at first by the slender runners, but these die in the following winter, if not before, and leave the plants as so many separate individuals.

92. **Tendrils** are branches of a very slender sort, like runners, not destined like them for propagation, and therefore always destitute

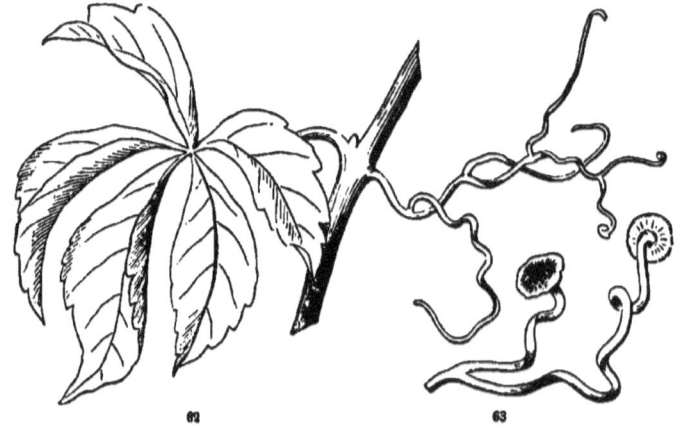

of buds or leaves, but intended for climbing. Those of the Grape-Vine, of the Virginia Creeper (Fig. 62), and of the Cucumber and

FIG. 62. Piece of the stem of Virginia Creeper, bearing a leaf and a tendril. 63. Tips of a tendril, about the natural size, showing the disks by which they hold fast to walls, &c.

Squash tribe are familiar illustrations. The tendril commonly grows straight and outstretched until it reaches some neighboring support, such as a stem, when its apex hooks around it to secure a hold; then the whole tendril shortens itself by coiling up spirally, and so draws the shoot of the growing plant nearer to the supporting object. When the Virginia Creeper climbs the side of a building or the smooth bark of a tree, which the tendrils cannot lay hold of in the usual way, their tips expand into a flat disk or sucker (Fig. 62, 63), which adheres very firmly to the wall or bark, enabling the plant to climb over and cover such a surface, as readily as the Ivy does by means of its sucker-like little rootlets. The same result is effected by different organs, in the one case by branches in the form of tendrils; in the other, by roots.

93. Tendrils, however, are not always branches; some are leaves, or parts of leaves, as those of the Pea (Fig. 20). Their nature in each case is to be learned from their position, whether it be that of a leaf or of a branch. In the same way

94. **Spines or Thorns** sometimes represent leaves, as in the Barberry, where their nature is shown by their situation *outside* of an axillary bud or branch. In other words, here they have a bud in their axil, and are therefore leaves; so we shall have to mention them in another place. Most commonly spines are stunted and hardened branches, arising from the axils of leaves, as in the Hawthorn and Pear. A neglected Pear-tree or Plum-tree shows every gradation between ordinary branches and thorns. Thorns sometimes branch, their branches partaking of the same spiny character: in this way those on the trunks of Honey-Locust trees (produced from adventitious buds, 58) become exceedingly complicated and horrid. The thorns on young shoots of the Honey-Locust may appear somewhat puzzling at first view; for they are situated some distance above the axil of the leaf. Here the thorn comes from the uppermost of several supernumerary buds (59). *Prickles*, such as those of the Rose and Blackberry, must not be confounded with thorns: these have not the nature of branches, and have no connection with the wood; but are only growths of the bark. When we strip off the bark, the prickles go with it.

95. Still stranger forms of stems and branches than any of these are met with in some tribes of plants, such as Cactuses (Fig. 76). These will be more readily understood after we have considered some of the commoner forms of

96. **Subterranean Stems and Branches.** These are very numerous and various ; but they are commonly overlooked, or else confounded with roots. From their situation they are out of the sight of the superficial observer : but if sought for and examined, they will well repay the student's attention. For the vegetation that is carried on under ground is hardly less varied, and no less interesting and important, than that which meets our view above ground. All their forms may be referred to four principal kinds ; namely, the *Rhizoma* or *Rootstock*, the *Tuber*, the *Corm*, and the *Bulb*.

97. **The Rootstock, or Rhizoma,** in its simplest form, is merely a creeping stem or branch (86) growing beneath the surface of the soil, or partly covered by it. Of this kind are the so-called *creeping, running,* or *scaly roots,* such as those by which the Mint (Fig. 64), the Scotch Rose, the Couch-grass or Quick-grass, and many other plants, spread so rapidly and widely, "by the root," as it is said.

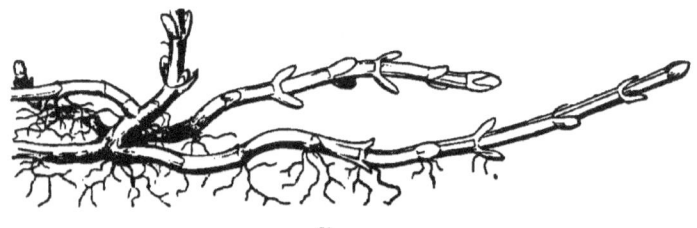

64

That these are really *stems,* and not roots, is evident from the way in which they grow ; from their consisting of a succession of joints ; and from the leaves which they bear on each joint (or *node,* as the botanist calls the place from which leaves arise), in the form of small scales, just like the lowest ones on the upright stem next the ground. Like other stems, they also produce buds in the axils of these scales, showing the scales to be leaves ; whereas real roots bear neither leaves nor axillary buds. Placed, as they are, in the damp and dark soil, such stems naturally produce roots, just as the creeping stem does where it lies on the surface of the ground ; but the whole appearance of these roots, their downward growth, and their mode of branching, are very different from that of the subterranean stem they spring from.

98. It is easy to see why plants with these running rootstocks take such rapid and wide possession of the soil, — often becoming great pests to farmers, — and why they are so hard to get rid of. They are

FIG. 64.   Rootstocks, or creeping subterranean branches, of the Peppermint.

always perennials (41) ; the subterranean shoots live over the first winter, if not longer, and are provided with vigorous buds at every joint. Some of these buds grow in spring into upright stems, bearing foliage, to elaborate the plant's crude food into nourishment, and at length produce blossoms for reproduction by seed ; while many others, fed by nourishment supplied from above, form a new generation of subterranean shoots ; and this is repeated over and over in the course of the season or in succeeding years. Meanwhile as the subterranean shoots increase in number, the older ones, connecting the series of generations into one body, die off year by year, liberating the already rooted side-branches as so many separate plants ; and so on indefinitely. Cutting these running rootstocks into pieces, therefore, by the hoe or the plough, far from destroying the plant, only accelerates the propagation ; it converts one many-branched plant into a great number of separate individuals. Even if you divide the shoots into as many pieces as there are joints of stem, each piece (Fig. 65) is already a plantlet, with its roots and with a bud in the axil of its scale-like leaf (either latent or apparent), and having prepared nourishment enough in the bit of stem to develop this bud into a leafy stem ; and so a single plant is all the more speedily converted into a multitude. Such plants as the Quick-grass accordingly realize the fable of the Hydra ; as fast as one of its many branches is cut off, twice as many, or more, spring up in its stead. Whereas, when the subterranean parts are only roots, cutting away the stem completely destroys the plant, except in the rather rare cases where the root produces adventitious buds (58).

65

99. The more nourishment rootstocks contain, the more readily do separate portions, furnished with buds, become independent plants. It is to such underground stems, thickened with a large amount of starch, or some similar nourishing matter stored up in their tissue, that the name of *rhizoma* or rootstock is commonly applied ; — such, for example, as those of the Sweet Flag or Calamus, of Ginger, of Iris or Flower-de-luce (Fig. 133), and of the Solomon's Seal (Fig. 66).

100. The rootstocks of the common sorts of Iris of the gardens usually lie on the surface of the ground, partly uncovered ; and they bear real leaves (Fig. 133), which closely overlap each other ;

FIG. 65. A piece of the running rootstock of the Peppermint, with its node or joint, and an axillary bud ready to grow.

A *

the joints (i. e. the *internodes*, or spaces between each leaf) being very short. As the leaves die, year by year, and decay, a scar left in the form of a ring marks the place where each leaf was attached. Instead of leaves, rootstocks buried under ground commonly bear scales, like those of the Mint (Fig. 64), which are imperfect leaves.

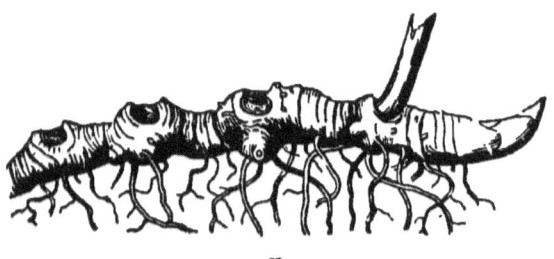

66

101. Some rootstocks are marked with large round scars of a different sort, like those of the Solomon's Seal (Fig. 66), which gave this name to the plant, from their looking something like the impression of a seal upon wax. Here the rootstock sends up every spring an herbaceous stalk or stem, which bears the foliage and flowers, and dies in autumn; and the *seal* is the circular scar left by the death and separation of the dead stalk from the living rootstock. As but one of these is formed each year, they mark the limits of a year's growth. The bud at the end of the rootstock in the figure, which was taken in summer, will grow the next spring into the stalk of the season, which, dying in autumn, will leave a similar scar, while another bud will be formed farther on, crowning the ever-advancing summit or growing end of the stem.

102. As each year's growth of stem, in all these cases, makes its own roots, it soon becomes independent of the older parts. And after a certain age, a portion dies off behind, every year, about as fast as it increases at the growing end; — death following life with equal and certain step, with only a narrow interval between. In vigorous plants of Solomon's Seal or Iris, the living rootstock is several inches or a foot in length; while in the short rootstock of

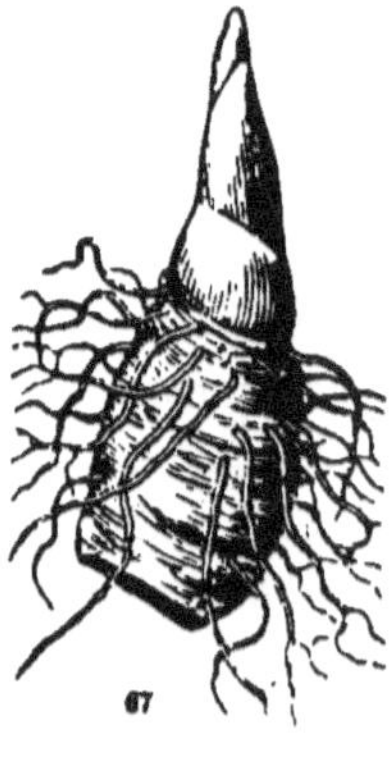

FIG. 66. Rootstock of Solomon's Seal, with the bottom of the stalk of the season, and the bud for the next year's growth.

FIG. 67. The very short rootstock and bud of a Trillium or Birthroot.

Trillium or Birthroot (Fig. 67) life is reduced to a very narrow span, only an inch or less intervening between death beneath and young life in the strong bud annually renewed at the summit.

103. **A Tuber** is a thickened portion of a rootstock. When slender subterranean branches, like those of the Quick-grass or Mint (Fig. 64), become enlarged at the growing end by the accumulation there of an abundance of solid nourishing matter, *tubers* are produced, like those of the Nut-grass of the Southern States (which accordingly becomes a greater pest even than the Quick-grass), and of the Jerusalem Artichoke, and the Potato. The whole formation may be seen at a glance in Figure 68, which represents the subterranean growth of a Potato-plant, and shows the tubers in all their stages, from shoots just beginning to enlarge at the tip, up to fully-formed potatoes. And Fig. 69, — one of the forming tubers moderately magnified, — plainly shows the leaves of this thickening shoot, in the form of little scales. It is under these scales that the *eyes* appear (Fig. 70) : and these are evidently axillary buds (43).

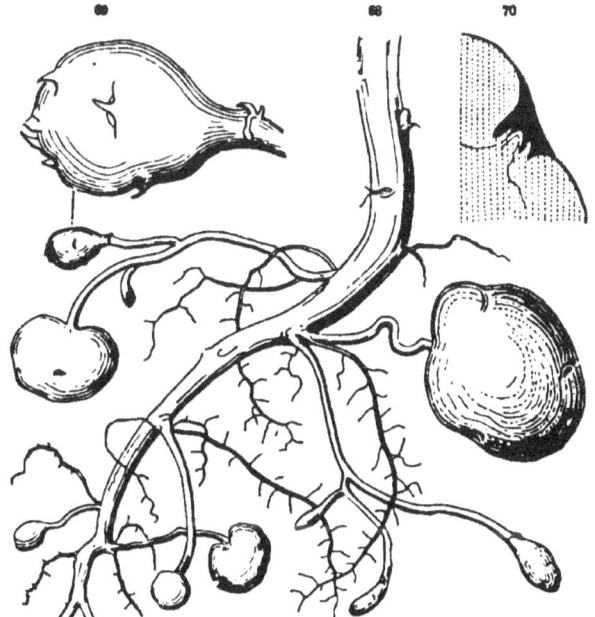

104. Let us glance for a moment at the economy or mode of life of the Potato-plant, and similar vegetables, as shown in the *mor-*

FIG. 68. Forming tubers of the Potato. 69. One of the very young potatoes, moderately magnified. 70. Slice of a portion through an eye, more magnified.

phology of the branches, — that is, in the different forms they appear under, and the purposes they serve. The Potato-plant has three principal forms of branches : — 1. Those that bear ordinary leaves, expanded in the air, to digest what they gather from it and what the roots gather from the soil, and convert it into nourishment. 2. After a while a second set of branches at the summit of the plant bear flowers, which form fruit and seed out of a portion of the nourishment which the leaves have prepared. 3. But a larger part of this nourishment, while in a liquid state, is carried down the stem, into a third sort of branches under ground, and accumulated in the form of starch at their extremities, which become tubers, or depositories of prepared solid food; — just as in the Turnip, Carro·, Dahlia, &c. (Fig. 57 – 60), it is deposited in the root. The use of the store of food is obvious enough. In the autumn the whole plant dies, except the seeds (if it formed them) and the tubers; and the latter are left disconnected in the ground. Just as that small portion of nourishing matter which is deposited in the seed (3, and Fig. 34) feeds the embryo when it germinates, so the much larger portion deposited in the tuber nourishes its buds, or eyes, when they likewise grow, the next spring, into new plants. And the great supply enables them to shoot with a greater vigor at the beginning, and to produce a greater amount of vegetation than the seedling plant could do in the same space of time ; which vegetation in turn may prepare and store up, in the course of a few weeks or months, the largest quantity of solid nourishing material, in a form most available for food. Taking advantage of this, man has transported the Potato from the cool Andes of South America to other cool climates, and makes it yield him a copious supply of food, especially in countries where the season is too short, or the summer's heat too little, for profitably cultivating the principal grain-plants.

105. All the sorts of subterranean stems or branches distinguished by botanists pass into one another by gradations. We have seen how nearly related the tuber is to the rootstock, and there are many cases in which it is difficult to say which is the proper name to use. So likewise,

106. The **Corm, or Solid Bulb,** like that of the Indian Turnip and the Crocus (Fig. 71), is just a very short and thick rootstock ; as will be seen by comparing Fig. 71 with Fig. 67. Indeed, it grows so very little in length, that it is often much broader than long, as in the Indian Turnip, and the Cyclamen of our greenhouses. Corms

are usually upright, producing buds on their upper surface and roots from the lower. But (as we see in the Crocus here figured) buds may shoot from just above any of the faint cross lines or rings, which are the scars left by the death  and decay of the sheathing bases of former leaves. That is, these are axillary buds. In these extraordinary (just as in ordinary) stems, the buds are either axillary or terminal. The whole mode of growth is just the same, only the corm does not increase in length faster than it does in thickness. After a few years some of the buds grow into new corms at the expense of the old one ; the young ones taking the nourishment from the parent, and storing up a large part of it in their own tissue.  When exhausted in this way, as well as by flowering, the old corm dies, and its shrivelled and decaying remains may be found at the side of or beneath the present generation, as we see in the Crocus (Fig. 71).

107. The corm of a Crocus is commonly covered with a thin and dry, scaly or fibrous husk, consisting of the dead remains of the bases of former leaves. When this husk consists of many scales, there is scarcely any distinction left between the corm and

108. The Bulb. This is an extremely short subterranean stem, usually much broader than high, producing roots from underneath, and covered with leaves or the bases of leaves, in the form of thickened scales. It is, therefore, the same as a corm, or solid bulb, only it bears an abundance of leaves or scales, which make up the greater part of its bulk. Or we may regard it as a bud, with thick and fleshy scales. Compare a Lily-bulb (Fig. 73) with the strong scaly buds of the Hickory and Horsechestnut (Fig. 48 and 49), and the resemblance will be apparent enough.

109. Bulbs serve the same purpose as tubers, rootstocks, or corms. The main difference is, that in these the store of food for future growth is deposited in the stem ; while in the bulb, the greater part is deposited in the bases of the leaves, changing them into thick scales, which closely overlap or enclose one another, because the stem does not elongate enough to separate them. That the scales

FIG. 71. Corm or solid bulb of a Crocus. 72. The same, cut through lengthwise.

of the bulb are the bases of leaves may be seen at once by following any of the ground-leaves (root-leaves as they are incorrectly

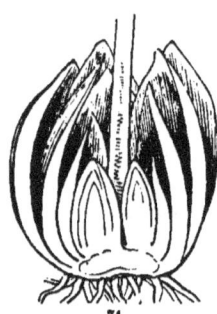

called) down to their
origin in the bulb.
Fig. 75 represents
one of them from
the White Lily; the
thickened base, which
makes a scale, being
cut off below, to show
its thickness. After
having lasted its time
and served its purpose as foliage, the green leaf dies, down to the
thickened base, which remains as a scale of the bulb. And year
after year, as the bulb grows from the centre, to produce the vegetation and the flowers of the season, the outer scales yield up their
store of nourishment for the purpose, and perish.

110. Each scale, being a leaf, may have a bud in its axil. Some
of these buds grow into leafy and flowering stems
above ground: others grow into new bulbs, feeding
on the parent, and at length destroying it, in the same
way that corms do, as just described (106).

111. When the scales are broad and enwrap all
that is within so as to form a succession of coats, one
over another, the bulb is said to be *tunicated* or *coated.*
The Tulip, Hyacinth, Leek, and Onion afford such
familiar examples of coated bulbs that no figure is
needed. When the scales are narrow and separate,
as in the Lily (Fig. 73), the bulb is said to be *scaly.*

112. **Bulblets** are small bulbs formed above ground
on some plants; as in the axils of the leaves of the
common bulbiferous Lily of the gardens, and often in
the flower-clusters of the Leek and Onion. They are
plainly nothing but bulbs with thickened scales. They
never grow into branches, but detach themselves when
full grown, and fall to the ground, to take root there and form
new plants.

113. From the few illustrations already given, attentive students

FIG. 73. Bulb of the Meadow or Canada Lily.   74. The same, cut through lengthwise.
FIG. 75. A lower leaf of White Lily, with its base under ground thickened into a bulb-
scale.

can hardly fail to obtain a good idea of what is meant by *morphology* in Botany; and they will be able to apply its simple principles for themselves to all forms of vegetation. They will find it very interesting to identify all these various subterranean forms with the common plan of vegetation above ground. There is the same structure, and the same mode of growth in reality, however different in appearance, and however changed the form, to suit particular conditions, or to accomplish particular ends. It is plain to see, already, that the plant is constructed *according to a plan*, — a very simple one, — which is exhibited by all vegetables, by the extraordinary no less than by the ordinary kinds; and that the same organ may appear under a great many different shapes, and fulfil very different offices.

114. These extraordinary shapes are not confined to subterranean vegetation. They are all repeated in various sorts of *fleshy plants;* in the Houseleek, Aloe, Agave (Fig. 82), and in the many and strange shapes which the Cactus family exhibit (Fig. 76); shapes which imitate rootstocks, tubers, corms, &c. above ground. All these we may regard as

115. **Consolidated Forms of Vegetation.** While ordinary plants are constructed on the plan of great spread of surface (131), these are formed on the plan of the least possible amount of surface in proportion to their bulk. The Cereus genus of Cactuses, for example, consisting of solid columnar trunks (Fig. 76, *b*), may be likened to rootstocks. A green rind serves the purpose of foliage ; but the surface is as nothing compared with an ordinary leafy plant of the same bulk. Compare, for instance, the largest Cactus known, the Giant Cereus of the Gila River (Fig. 76, in the background), which rises to the height of fifty or sixty feet, with a common leafy tree of the same height, such as that in Fig. 54, and estimate how vastly greater, even without the foliage, the surface of the latter is than that of the former. Compare, in the same view, an Opuntia or Prickly-Pear Cactus, its stem and branches formed of a succession of thick and flattened joints (Fig. 76, *a*), which may be likened to tubers, or an Epiphyllum (*d*), with shorter and flatter joints, with an ordinary leafy shrub or herb of equal size. And finally, in Melon-Cactuses or Echinocactus (*c*), with their globular or bulb-like shapes, we have plants in the compactest shape ; their spherical figure being such as to expose the least possible amount of its bulk to the air.

116. These *consolidated plants* are evidently adapted and designed

for very *dry regions ;* and in such only are they found.   Similarly,
bulbous and corm-bearing plants, and the like, are examples of a
form of vegetation which in the growing season may expand a large
surface to the air and light, while during the period of rest the
living vegetable is reduced to a globe, or solid form of the least
possible surface ; and this is protected by its outer coats of dead
and dry scales, as well as by its situation under ground.   Such
plants exhibit another and very similar adaptation to a season of
drought.   And they mainly belong to countries (such as Southern
Africa, and parts of the interior of Oregon and California) which
have a long hot season during which little or no rain falls, when,
their stalks and foliage above and their roots beneath being early cut
off by drought, the plants rest securely in their compact bulbs, filled
with nourishment, and retaining their moisture with great tenacity,
until the rainy season comes round.   Then they shoot forth leaves
and flowers with wonderful rapidity, and what was perhaps a desert
of arid sand becomes green with foliage and gay with blossoms,
almost in a day.   This will be more perfectly understood when the
nature and use of foliage have been more fully considered.   (Fig. 76
represents several forms of Cactus vegetation.)

a   b   c   d

76

LESSON VII.

MORPHOLOGY OF LEAVES.

117. In describing the subterranean forms of the stem, we have been led to notice already some of the remarkable forms under which leaves occur; namely, as *scales*, sometimes small and thin, as those of the rootstocks of the Quick-grass, or the Mint (Fig. 64), sometimes large and thick, as those of bulbs (Fig. 73 – 75), where they are commonly larger than the stem they belong to. We have seen, too, in the second Lesson, the seed-leaves (or cotyledons) in forms as unlike foliage as possible ; and in the third Lesson we have spoken of bud-scales as a sort of leaves. So that the botanist recognizes the leaf under other forms than that of foliage.

118. We may call foliage the *natural form* of leaves, and look upon the other sorts as *special forms*, — as *transformed* leaves : by this term meaning only that what would have been ordinary leaves under other circumstances (as, for instance, those on shoots of Mint, Fig. 64, had these grown upright in the air, instead of creeping under ground) are developed in special forms to serve some particular purpose. For the Great Author of Nature, having designed plants upon one simple plan, just adapts this plan to all cases. So, whenever any special purpose is to be accomplished, no new instruments or organs are created for it, but one of the three general organs of the vegetable, *root*, *stem*, or *leaf*, is made to serve the purpose, and is adapted to it by taking some peculiar form.

119. It is the study of the varied forms under this view that constitutes *Morphology* (61), and gives to this part of Botany such great interest. We have already seen stems and roots under a great variety of forms. But leaves appear under more various and widely different forms, and answer a greater variety of purposes, than do both the other organs of the plant put together. We have to consider, then, *leaves as foliage*, and *leaves as something else than foliage*. As we have just been noticing cases of leaves that are not foliage, we may consider these first, and enumerate the principal kinds.

120. **Leaves as Depositories of Food.** Of these we have had plenty of instances in the seed-leaves, such as those of the Almond, Apple-

seed (Fig. 11), Beech (Fig. 13 – 15), the Bean and Pea (Fig. 16 – 20), the Oak (Fig. 21, 22), and Horsechestnut (Fig. 23, 24) ; where the food upon which the plantlet feeds when it springs from the seed is stored up in its cotyledons or first leaves. And we have noticed how very unlike foliage such leaves are. Yet in some cases,

as in the Pumpkin (Fig. 10), they actually grow into green leaves as they get rid of their burden.

121. **Bulb-Scales** (Fig. 73 – 75) offer another instance, which we were considering at the close of the last Lesson. Here a part of the nourishment prepared in the foliage of one year is stored up in the scales, or subterranean thickened leaves, for the early growth and flowering of the next year ; and this enables the flowers to appear before the leaves, or as soon as they do ; as in Hyacinths, Snowdrops, and many bulbous plants.

122. **Leaves as Bud-scales, &c.** True to its nature, the stem produces leaves even under ground, where they cannot serve as foliage, and where often, as on rootstocks and tubers (97 – 103), they are not of any use that we know of. In such cases they usually appear as thin scales. So the first leaves of the stems of herbs, as they sprout from the ground, are generally mere scales, such as those of an Asparagus shoot ; and such are the first leaves on the stem of the seedling Oak (Fig. 22) and the Pea (Fig. 20). Similar scales, however, often serve an important purpose ; as when they form the covering of buds, where they protect the tender parts within (44). That bud-scales are

77

FIG. 77. Leaves of a developing bud of the Low Sweet Buckeye (Æsculus parviflora), showing a nearly complete set of gradations from a scale to a compound leaf of five leaflets.

leaves is plainly shown, in many cases, by the gradual transition between them and the first foliage of the shoot. The Common Lilac and the Shell-bark Hickory are good instances of the sort. But the best illustration is furnished by the Low Sweet Buckeye of the Southern States, which is often cultivated as an ornamental shrub. From one and the same growing bud we may often find all the gradations which are shown in Fig. 77.

123. **Leaves as Spines** occur in several plants. The most familiar instance is that of the Common Barberry. In almost any summer shoot, most of the gradations may be seen between the ordinary leaves, with sharp bristly teeth, and leaves which are reduced to a branching spine or thorn, as shown in Fig. 78. The fact that the spines of the Barberry produce a leaf-bud in their axil also proves them to be leaves.

124. **Leaves as Tendrils** are to be seen in the Pea and the Vetch (Fig. 20, 127), where the upper part of each leaf becomes a tendril, which the plant uses to climb by; and in one kind of Vetch the whole leaf is such a tendril.

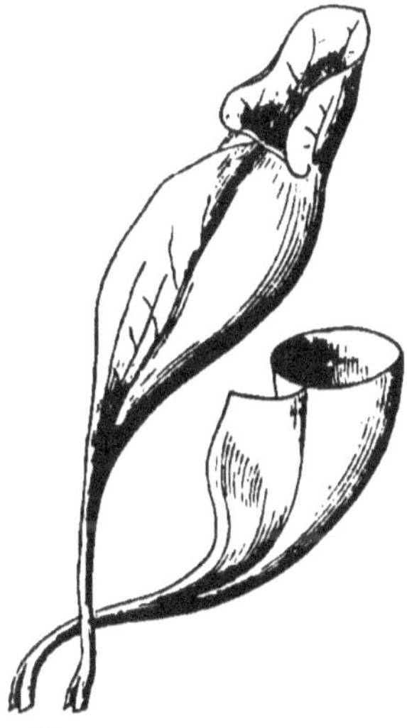

125. **Leaves as Pitchers**, or hollow tubes, are familiar to us in the common Pitcher-plant or Side-saddle Flower (Sarracenia, Fig. 79) of our bogs. These pitchers are generally half-full of water, in which flies and other insects are drowned, often in such numbers as to make a rich manure for the plant, no doubt; though we can hardly imagine this to be the design of the pitcher. Nor do we perceive here any need of a contrivance to hold water, since the roots of these plants are always well supplied by the wet bogs where they grow.

FIG. 78. Summer shoot of Barberry, showing the transition of leaves into spines.
FIG. 79. Leaf of Sarracenia purpurea, entire, and another with the upper part cut off.

126. Leaves as Fly-traps. Insects are caught in another way, and more expertly, by the most extraordinary of all the plants of this

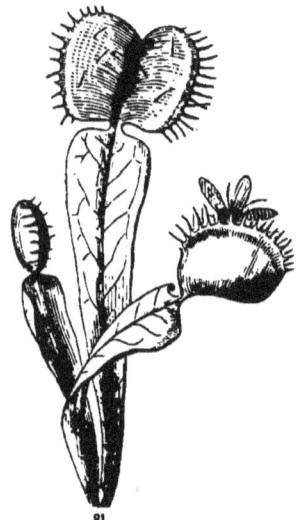

country, the Dionæa or Venus's Fly-trap, which grows in the sandy bogs around Wilmington, North Carolina. Here (Fig. 81) each leaf bears at its summit an appendage which opens and shuts, in shape something like a steel-trap, and operating much like one. For when open, as it commonly is when the sun shines, no sooner does a fly alight on its surface, and brush against any one of the several long bristles that grow there, than the trap suddenly closes, often capturing the intruder, pressing it all the harder for its struggles, and commonly depriving it of life. If the fly escapes, the trap soon slowly opens, and is ready for another capture. When retained, the insect is after a time moistened by a secretion from minute glands of the inner surface, and is apparently digested! How such and various other movements are made by plants, — some as quick as in this case, others very slow, but equally wonderful, — must be considered in a future Lesson.

127. Leaves serving both Ordinary and Special Purposes. Let us now remark, that the same leaf frequently answers its general purpose, as foliage, and some special purpose besides. For example, in the Dionæa, the lower part of the leaf, and probably the whole of it, acts as foliage, while the appendage serves its mysterious purpose as a fly-catcher. In the Pea and Vetch (Fig. 20, 127), the lower part of the leaf

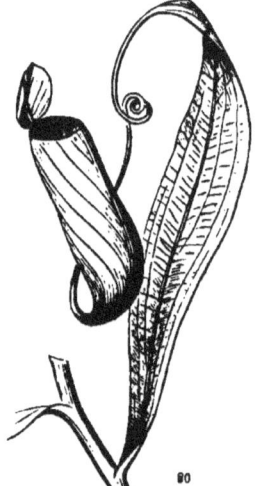

is foliage, the upper a tendril. In the Pitcher-plants of the Indian Archipelago (Nepenthes, Fig. 80) which are not rare in conservatories, the lower part of the leaf is expanded and acts as foliage;

FIG. 80. Leaf of Nepenthes: leaf, tendril, and pitcher combined.
FIG. 81. Leaves of Dionæa; the trap in one of them open, in the others closed.

farther on, it is contracted into a tendril, enabling the plant to climb; the end of this tendril is then expanded into a pitcher, of five or six inches in length, and on the end of this is a lid, which exactly closes the mouth of the pitcher until after it is full grown, when the lid opens by a hinge! But the whole is only one leaf.

128. So in the root-leaves of the Tulip or the Lily (Fig. 75), while the green leaf is preparing nourishment throughout the growing season, its base under ground is thickened into a reservoir for storing up a good part of the nourishment for next year's use.

129. Finally, the whole leaf often serves both as foliage, to prepare nourishment, and as a depository to store it up. This takes place in all fleshy-leaved plants, such as the Houseleek, the Ice-plant, and various sorts of Mesembryanthemum, in the Live-for-ever of the gardens to some extent, and very strikingly in the Aloe, and in the Century-plant. In the latter it is only the green surface of these large and thick leaves (of three to five feet in length on a strong plant, and often three to six inches thick near the base) which acts as foliage; the whole interior is white, like the interior of a potato, and almost as heavily loaded with starch and other nourishing matter. (Fig. 82 represents a young Century-plant, Agave Americana.)

LESSON VIII.

MORPHOLOGY OF LEAVES AS FOLIAGE.

130. HAVING in the last Lesson glanced at some of the special or extraordinary forms and uses of leaves, we now return to leaves in their ordinary condition, namely, as foliage. We regard this as the natural state of leaves. For although they may be turned to account in other and very various ways, as we have just seen, still their proper office in vegetation is to serve as foliage. In this view we may regard

131. **Leaves as a Contrivance for Increasing the Surface** of that large part of the plant which is exposed to the light and the air. This is shown by their expanded form, and ordinarily slight thickness in comparison with their length and breath. While a Melon-Cactus (115, Fig. 76) is a striking example of a plant with the least possible amount of surface for its bulk, a repeatedly branching leafy herb or tree presents the largest possible extent of surface to the air. The actual amount of surface presented by a tree in full leaf is much larger than one would be apt to suppose. Thus, the Washington Elm at Cambridge — a tree of no extraordinary size — was some years ago estimated to produce a crop of seven millions of leaves, exposing a surface of 200,000 square feet, or about five acres, of foliage.

132. What is done by the foliage we shall have to explain in another place. Under the present head we are to consider ordinary leaves as to their *parts* and their *shapes*.

133. **The Parts of the Leaf.** The principal part of a leaf is the blade, or expanded portion, one face of which naturally looks toward the sky, the other towards the earth. The blade is often raised on a stalk of its own, and on each side of the stalk at its base there is sometimes an appendage called a *stipule*. A complete leaf, therefore consists of a *blade* (Fig. 83, *b*), a *foot-stalk* or *leaf-stalk*, called the *petiole* (*p*), and a pair of *stipules* (*st*). See also Fig. 136.

134. It is the blade which we are now to describe. This, as being the essential and conspicuous part, we generally regard as the leaf: and it is only when we have to particularize, that we speak of the *blade*, or *lamina*, of the leaf.

135. Without here entering upon the subject of the anatomy of the leaf, we may remark, that leaves consist of two sorts of material, viz.: 1. the *green pulp*, or *parenchyma*; and 2. the *fibrous framework*, or skeleton, which extends throughout the soft green pulp and supports it, giving the leaf a strength and firmness which it would not otherwise possess. Besides, the whole surface is covered with a transparent skin, called the *epidermis*, like that which covers the surface of the shoots, &c.

136. The framework consists of *wood*, — a fibrous and tough material which runs from the stem through the leaf-stalk, when there is one, in the form of parallel threads or bundles of fibres; and in the blade these spread out in a horizontal direction, to form the *ribs* and *veins* of the leaf. The stout main branches of the framework (like those in Fig. 50) are called the *ribs*. When there is only one, as in Fig. 83, &c., or a middle one decidedly larger than the rest, it is called the *midrib*. The smaller divisions are termed *veins*; and their still smaller subdivisions, *veinlets*.

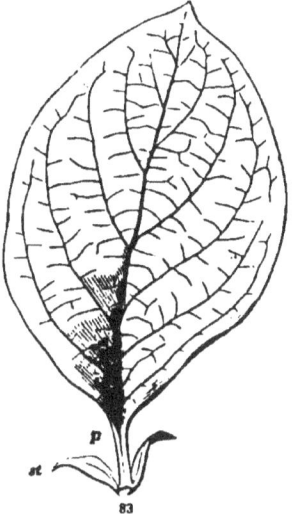

137. The latter subdivide again and again, until they become so fine that they are invisible to the naked eye. The fibres of which they are composed are hollow; forming tubes by which the sap is brought into the leaves and carried to every part. The arrangement of the framework in the blade is termed the

138. **Venation**, or mode of veining. This corresponds so completely with the general shape of the leaf, and with the kind of division when the blade is divided or lobed, that the readiest way to study and arrange the forms of leaves is first to consider their veining.

139. Various as it appears in different leaves, the veining is all reducible to two principal kinds; namely, the *parallel-veined* and the *netted-veined*.

140. In *netted-veined* (also called *reticulated*) leaves, the veins branch off from the main rib or ribs, divide into finer and finer

FIG. 83. Leaf of the Quince: *b*, blade; *p*, petiole; *st*, stipules.

veinlets, and the branches unite with each other to form meshes of network. That is, they *anastomose*, as anatomists say of the veins and arteries of the body. The Quince-leaf, in Fig. 83, shows this kind of veining in a leaf with a single rib. The Maple, Basswood, and Buttonwood (Fig. 50) show it in leaves of several ribs.

141. In *parallel-veined* leaves, the whole framework consists of slender ribs or veins, which run parallel with each other, or nearly so, from the base to the point of the leaf, not dividing and subdividing, nor forming meshes, except by very minute cross-veinlets. The leaf of any grass, or that of the Lily of the Valley (Fig. 84) will furnish a good illustration.

142. Such simple, parallel veins Linnæus, to distinguish them.

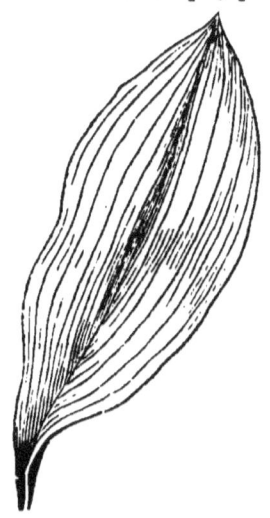

called *nerves*, and parallel-veined leaves are still commonly called *nerved* leaves, while those of the other kind are said to be *veined;* — terms which it is convenient to use, although these " nerves " and " veins " are all the same thing, and have no likeness to the *nerves* of animals.

143. *Netted-veined* leaves belong to plants which have a pair of seed-leaves or cotyledons, such as the Maple (Fig. 1 –7), Beech (Fig. 15), Pea and Bear. (Fig. 18, 20), and most of the illustrations in the first and second Lessons. While *parallel-veined* or *nerved* leaves belong to plants with one cotyledon or true seed-leaf; such as the Iris (Fig. 134) and Indian Corn (Fig. 42). So that a mere glance at the leaves of the tree or herb enables one to tell what the structure of the embryo is, and to refer the plant to one or the other of these two grand classes, — which is a great convenience. For generally when plants differ from each other in some one important respect, they differ correspondingly in other respects as well.

144. Parallel-veined leaves are of two sorts; one kind, and the commonest, having the ribs or nerves all running from the base to the point of the leaf, as in the examples already given; while in another kind they run from a midrib to the margin; as in the com-

FIG. 84. A (parallel-veined) leaf of the Lily of the Valley.

mon Pickerel-weed of our ponds, in the Banana (Fig. 47), and many similar plants of warm climates.

145. Netted-veined leaves are also of two sorts, as is shown in the examples already referred to. In one case the veins all rise from a single rib (the midrib), as in Fig. 83. Such leaves are called *feather-veined* or *pinnately-veined;* both terms meaning the same thing, namely, that the veins are arranged on the sides of the rib like the plume of a feather on each side of the shaft.

146. In the other case (as in the Buttonwood, Fig. 50, Maple, &c.), the veins branch off from three, five, seven, or nine ribs, which spread from the top of the leaf-stalk, and run through the blade like the toes of a web-footed bird. Hence these are said to be *palmately* or *digitately* veined, or (since the ribs diverge like rays from a centre) *radiate-veined.*

147. Since the general outline of leaves accords with the framework or skeleton, it is plain that *feather-veined* leaves will incline to elongated shapes, or at least will be longer than broad ; while in *radiate-veined* leaves more rounded forms are to be expected. A glance at the following figures shows this. Whether we consider the veins of the leaf to be adapted to the shape of the blade, or the green pulp to be moulded to the framework, is not very material. Either way, the outline of each leaf corresponds with the mode of spreading, the extent, and the relative length of the veins. Thus, in oblong or elliptical leaves of the feather-veined sort (Fig. 87, 88), the principal veins are nearly equal in length ; while in ovate and heart-shaped leaves (Fig. 89, 90), those below the middle are longest ; and in leaves which widen upwards (Fig. 91 – 94), the veins above the middle are longer than the others.

148. Let us pass on, without particular reference to the kind of veining, to enumerate the principal

149. **Forms of Leaves as to General Outline.** It is necessary to give names to the principal shapes, and to define them rather precisely, since they afford the easiest marks for distinguishing species. The same terms are used for all other flattened parts as well, such as the petals of the flowers ; so that they make up a great part of the descriptive language of Botany. We do not mention the names of common plants which exhibit these various shapes. It will be a good exercise for young students to look them up and apply them.

150. Beginning with the narrower and proceeding to the broadest forms. a leaf is said to be

S & F—4

Linear (Fig. 85), when narrow, several times longer than wide, and of the same breadth throughout.

Lanceolate, or *lance-shaped*, when several times longer than wide, and tapering upwards (Fig. 86), or both upwards and downwards.

Oblong (Fig. 87), when nearly twice or thrice as long as broad.

Elliptical (Fig. 88) is oblong with a flowing outline, the two ends alike in width.

Oval is the same as broadly elliptical, or elliptical with the breadth considerably more than half the length.

Ovate (Fig. 89), when the outline is like a section of a hen's-egg lengthwise, the broader end downward.

Orbicular, or *rotund* (Fig. 102), circular in outline, or nearly so.

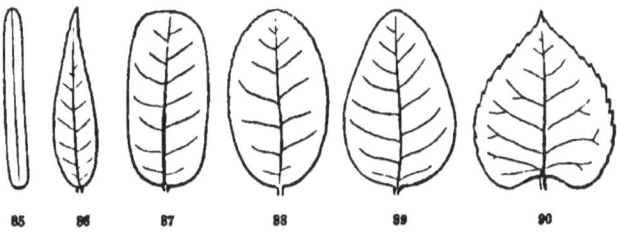

85   86   87   88   89   90

151. When the leaf tapers towards the base, instead of upwards, it may be

Oblanceolate (Fig. 91), which is lance-shaped, with the more tapering end downwards;

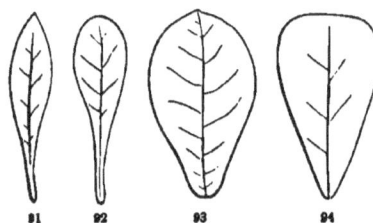

Spatulate (Fig. 92), rounded above and long and narrow below, like a spatula;

Obovate (Fig. 93), or inversely ovate, that is, ovate with the narrower end down; or

91   92   93   94

Cuneate, or *cuneiform*, that is, *wedge-shaped* (Fig. 94), broad above and tapering by straight lines to an acute angle at the base.

152. As to the Base, its shape characterizes several forms, such as

Cordate, or *heart-shaped* (Fig. 90, 99, 8), when a leaf of an ovate form, or something like it, has the outline of its rounded base turned in (forming a notch or *sinus*) where the stalk is attached.

Reniform, or *kidney-shaped* (Fig. 100), like the last, only rounder and broader than long.

FIG. 85 - 90.   Various forms of feather-veined leaves.
FIG. 91.   Oblanceolate, 92. spatulate, 93. obovate, 94. wedge-shaped, feather-veined leaves.

Auriculate, or *eared*, having a pair of small and blunt projections, or *ears*, at the base, as in one species of Magnolia (Fig. 96).

Sagittate, or *arrow-shaped*, where such ears are pointed and turned downwards, while the main body of the blade tapers upwards to a point, as in the common Sagittaria or Arrow-head, and in the Arrow-leaved Polygonum (Fig. 95).

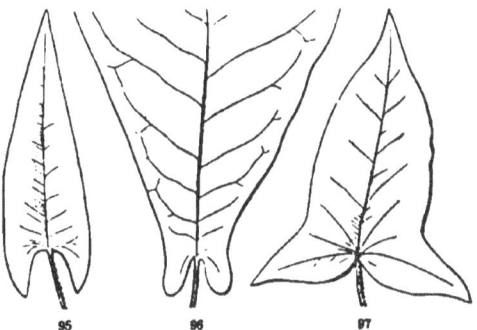

Hastate, or *halberd-shaped*, when such lobes at the base point outwards, giving the leaf the shape of the halberd of the olden time, as in another Polygonum (Fig. 97).

Peltate, or *shield-shaped*, (Fig. 102,) is the name applied to a curious modification of the leaf, commonly of a rounded form, where the footstalk is attached to the lower surface, instead of the base, and

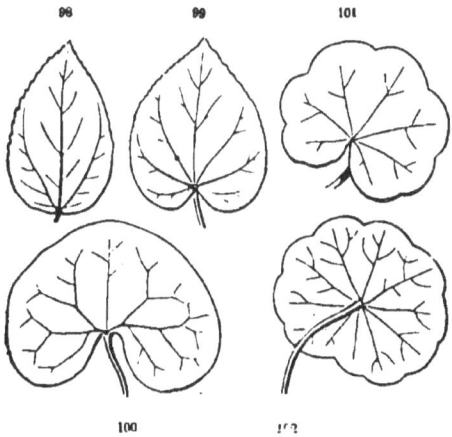

therefore is naturally likened to a shield borne by the outstretched arm. The common Watershield, the Nelumbium, and the White Water-lily, and also the Mandrake, exhibit this sort of leaf. On comparing the shield-shaped leaf of the common Marsh Pennywort (Fig. 102) with that of another common species (Fig. 101), we see at once what this peculiarity means. A shield-shaped leaf is like a

FIG. 95. Sagittate, 96. auriculate, 97. halberd-shaped, leaves.
FIG. 98 – 102. Various forms of radiate-veined leaves.

kidney-shaped (Fig. 100) or other rounded leaf, with the margins at the base brought together and united.

153. **As to the Apex,** the following terms express the principal variations.

Acuminate, pointed, or *taper-pointed,* when the summit is more or less prolonged into a narrowed or tapering point, as in Fig. 97.

Acute, when ending in an acute angle or not prolonged point, as in Fig. 104, 98, 95, &c.

Obtuse, when with a blunt or rounded point, as in Fig. 105, 89, &c.

Truncate, with the end as if cut off square, as in Fig. 106, 94.

Retuse, with the rounded summit slightly indented, forming a very shallow notch, as in Fig. 107.

Emarginate, or *notched,* indented at the end more decidedly, as in Fig. 108.

Obcordate, that is, inversely heart-shaped, where an obovate leaf is more deeply notched at the end (Fig. 109), as in White Clover and Wood-sorrel ; so as to resemble a cordate leaf (Fig. 99) inverted.

Cuspidate, tipped with a sharp and rigid point ; as in Fig. 110.

Mucronate, abruptly tipped with a small and short point, like a projection of the midrib ; as in Fig. 111.

Aristate, awn-pointed, and *bristle-pointed,* are terms used when this mucronate point is extended into a -longer bristle-form or other slender appendage.

The first six of these terms can be applied to the lower as well as to the upper end of a leaf or other organ. The others belong to the apex only.

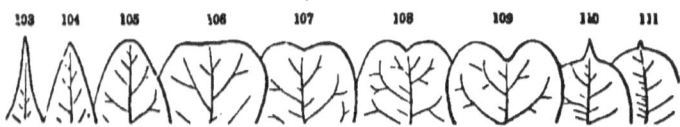

103   104   105   106   107   108   109   110   111

FIG. 103 - 111. Forms of the apex of leaves.

LESSON IX.

MORPHOLOGY OF LEAVES AS FOLIAGE. — SIMPLE AND COM-
POUND LEAVES, STIPULES, ETC.

154. In the foregoing Lesson leaves have been treated of in their simplest form, namely, as consisting of a single blade. But in many cases the leaf is divided into a number of separate blades. That is,

155. **Leaves are either Simple or Compound.** They are said to be *simple*, when the blade is all of one piece : they are *compound*, when the blade consists of two or more separate pieces, borne upon a common leaf-stalk. And between these two kinds every intermediate gradation is to be met with. This will appear as we proceed to notice the principal

156. **Forms of Leaves as to particular Outline** or degree of division. In this respect, leaves are said to be

Entire, when their general outline is completely filled out, so that the margin is an even line, without any teeth or notches ; as in Fig. 83, 84, 100, &c.

Serrate, or *saw-toothed*, when the margin only is cut into sharp teeth, like those of a saw, and pointing forwards; as in Fig. 112; also 90, &c.

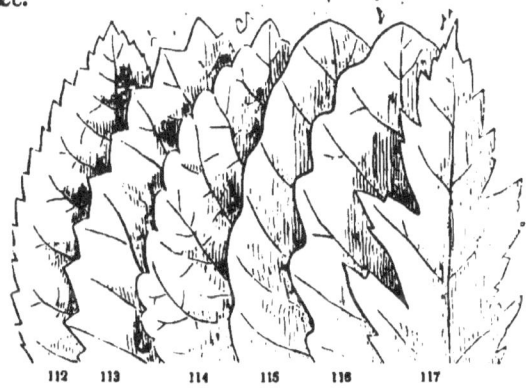

112   113   114   115   116   117

Dentate, or *toothed*, when such teeth point outwards, instead of forwards ; as in Fig. 113.

FIG. 112–117. Kinds of margin of leaves.

6

Crenate, or *scalloped*, when the teeth are broad and rounded ; as in Fig. 114, 101.

Repand, *undulate*, or *wavy*, when the margin of the leaf forms a wavy line, bending slightly inwards and outwards in succession ; as in Fig. 115.

Sinuate, when the margin is more strongly sinuous, or turned inwards and outwards, as in Fig. 116.

Incised, *cut*, or *jagged*, when the margin is cut into sharp, deep, and irregular teeth or incisions, as in Fig. 117.

157. When leaves are more deeply cut, and with a definite number of incisions, they are said, as a general term, to be *lobed ;* the parts being called *lobes.* Their number is expressed by the phrase *two-lobed, three-lobed, five-lobed, many-lobed,* &c., as the case may be. When the depth and character of the lobing needs to be more particularly specified, — as is often the case, — the following terms are employed, viz. :

Lobed, when the incisions do not extend deeper than about half-way between the margin and the centre of the blade, if so far, and are more or less rounded ; as in the leaves of the Post-Oak, Fig. 118, and the Hepatica, Fig. 122.

Cleft, when the incisions extend half-way down or more, and especially when they are sharp, as in Fig. 119, 123. And the phrases *two-cleft,* or, in the Latin form, *bifid ; three-cleft,* or *trifid ; four-cleft,* or *quadrifid ; five-cleft,* or *quinquefid,* &c.; or *many-cleft,* in the Latin form *multifid,* — express the number of the *segments,* or portions.

Parted, when the incisions are still deeper, but yet do not quite reach to the midrib or the base of the blade ; as in Fig. 120, 124. And the terms *two-parted, three-parted,* &c. express the number of such divisions.

Divided, when the incisions extend quite to the midrib, as in the lower part of Fig. 121 ; or to the leaf-stalk, as in Fig. 125 ; which makes the leaf compound. Here, using the Latin form, the leaf is said to be *bisected, trisected* (Fig. 125), &c., to express the number of the divisions.

158. In this way the *degree* of division is described. We may likewise express the *mode* of division. The notches or incisions, being places where the green pulp of the blade has not wholly filled up the framework, correspond with the veining ; as we perceive on comparing the figures 118 to 121 with figures 122 to 125. The

upper row of figures consists of *feather-veined*, or, in Latin form, *pinnately-veined* leaves (145); the lower row, of *radiate-veined* or *palmately-veined* leaves (146).

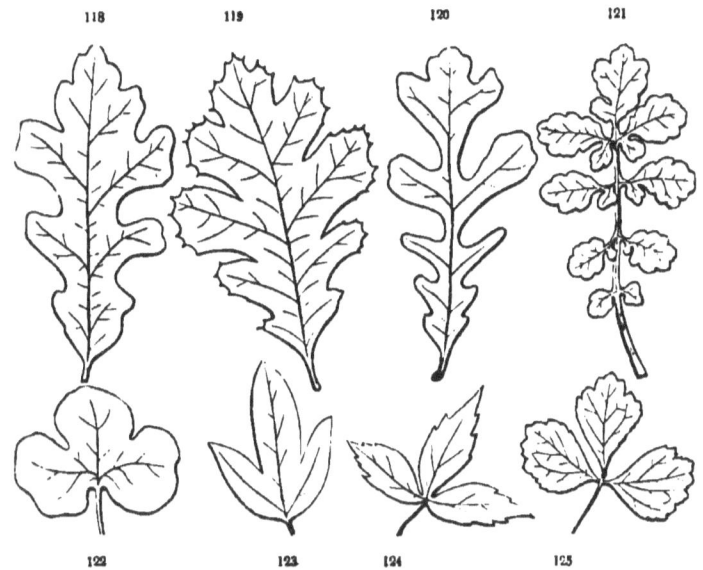

159. In the upper row the incisions all point towards the midrib, from which the main veins arise, the incisions (or *sinuses*) being between the main veins. That is, being *pinnately* veined, such leaves are *pinnately lobed* (Fig. 118), *pinnately cleft*, or *pinnatifid* (Fig. 119), *pinnately parted* (Fig. 120), or *pinnately divided* (Fig. 121), according to the depth of the incisions, as just defined.

160. In the lower row of figures, as the main veins or ribs all proceed from the base of the blade or the summit of the leaf-stalk, so the incisions all point in that direction. That is, *palmately*-veined leaves are *palmately lobed* (Fig. 122), *palmately cleft* (Fig. 123), *palmately parted* (Fig. 124), or *palmately divided* (Fig. 125). Sometimes, instead of palmately, we say *digitately* cleft, &c., which means just the same.

161. To be still more particular, the number of the lobes, &c. may come into the phrase. Thus, Fig. 122 is a *palmately three-lobed*; Fig. 123, a *palmately three-cleft*; Fig. 124, a *palmately three-parted*; Fig. 125, a *palmately three-divided*, or *trisected*, leaf. The

FIG. 118–121. Pinnately lobed, cleft, parted, and divided leaves.
FIG. 122–125. Palmately or digitately lobed, cleft, parted, and divided leaves.

Sugar-Maple and the Buttonwood (Fig. 50) have *palmately five-lobed leaves;* the Soft White-Maple *palmately five-parted leaves;* and so on.   And in the other sort, the Post-Oak has *pinnately seven- to nine-lobed leaves;* the Red-Oak commonly has *pinnately seven- to nine-cleft leaves,* &c., &c.

162. The divisions, lobes, &c. may themselves be *entire* (without teeth or notches, 156), as in Fig. 118, 122, &c.; or *serrate* (Fig. 124), or otherwise toothed or incised (Fig. 121); or else lobed, cleft, parted, &c.: in the latter cases making *twice pinnatifid, twice palmately* or *pinnately lobed, parted,* or *divided* leaves, &c.   From these illustrations, the student will perceive the plan by which the botanist, in two or three words, may describe any one of the almost endlessly diversified shapes of leaves, so as to convey a perfectly clear and definite idea of it.

163. **Compound Leaves.**   These, as already stated (155), do not differ in any absolute way from the *divided* form of simple leaves. A compound leaf is one which has its blade in two or more entirely separate parts, each usually with a stalklet of its own: and the stalklet is often *jointed* (or *articulated*) with the main leaf-stalk, just as this is jointed with the stem.   When this is the case, there is no

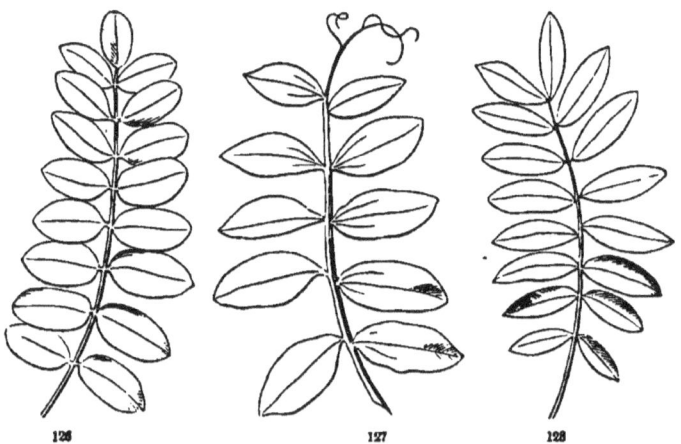

126   127   128

doubt that the leaf is compound.   But when the pieces have no stalklets, and are not jointed with the main leaf-stalk, the leaf may be considered either as simple and divided, or compound, according to the circumstances.

FIG. 126. Pinnate with an odd leaflet, or odd-pinnate.   127. Pinnate with a tendril 126. Abruptly pinnate leaf.

164. The separate pieces or little blades of a compound leaf are called *leaflets*.

165. Compound leaves are of two principal kinds, namely, the *pinnate* and the *palmate ;* answering to the two modes of veining in reticulated leaves (145 – 147), and to the two sorts of lobed or divided leaves (158, 159).

166. *Pinnate* leaves are those in which the leaflets are arranged on the sides of a main leaf-stalk ; as in Fig. 126 – 128. They answer to the *feather-veined* (i. e. *pinnately-veined*) simple leaf ; as will be seen at once, on comparing Fig. 126 with the figures 118 to 121. The *leaflets* of the former answer to the *lobes* or *divisions* of the latter ; and the continuation of the petiole, along which the leaflets are arranged, answers to the midrib of the simple leaf.

167. Three sorts of pinnate leaves are here given. Fig. 126 is *pinnate with an odd or end leaflet*, as in the Common Locust and the Ash. Fig. 127 is *pinnate with a tendril at the end*, in place of the odd leaflet, as in the Vetches and the Pea. Fig. 128 is *abruptly pinnate*, having a pair of leaflets at the end, like the rest of the leaflets ; as in the Honey-Locust.

168. *Palmate* (also named *digitate*) leaves are those in which the leaflets are all borne on the very tip of the leaf-stalk, as in the Lupine, the Common Clover (Fig. 136), the Virginia Creeper (Fig. 62), and the Horsechestnut and Buckeye (Fig. 129). They answer to the *radiate-veined* or *palmately-veined* simple leaf ; as is seen by comparing Fig. 136 with the figures 122 to 125. That is, the Clover-leaf of three leaflets is the same as a palmately three-ribbed leaf cut into three separate leaflets. And such a simple five-lobed leaf as that of the Sugar-Maple, if more cut, so as to separate the parts, would produce a palmate leaf of five leaflets, like that of the Horsechestnut or Buckeye (Fig. 129).

129

169. Either sort of compound leaf may have any number of leaflets ; though palmate leaves cannot well have a great many, since they are all crowded together on the end of the main leaf-stalk.

FIG. 129. Palmate leaf of five leaflets, of the Sweet Buckeye.

Some Lupines have nine or eleven ; the Horsechestnut has seven, the Sweet Buckeye more commonly five, the Clover three. A pinnate leaf often has only seven or five leaflets, as in the Wild Bean or Groundnut ; and in the Common Bean it has only three ; in

some rarer cases only two ; in the Orange and Lemon only one! The joint at the place where the leaflet is united with the petiole alone distinguishes this last case from a simple leaf.*

170. The leaflets of a compound leaf may be either *entire* (as in Fig. 126 – 128), or *serrate*, or lobed, cleft, parted, &c. : in fact, they may present all the variations of simple leaves, and the same terms equally apply to them.

171. When this division is carried so far as to separate what would be one leaflet into two, three, or several, the leaf becomes *doubly* or *twice compound*, either *pinnately* or *palmately*, as the case may be.

130

For example, while some of the leaves of the Honey-Locust are *simply pinnate*, that is, *once pinnate*, as in Fig. 128, the greater part

---

* When the botanist, in describing leaves, wishes to express the number of leaflets, he may use terms like these : —

Unifoliolate, for a compound leaf of a single leaflet ; from the Latin *unum*, one, and *foliolum*, leaflet.

Bifoliolate, of two leaflets, from the Latin *bis*, twice, and *foliolum*, leaflet.

Trifoliolate (or *ternate*), of three leaflets, as the Clover ; and so on.

When he would express in one phrase both the number of leaflets and the way the leaf is compound, he writes : —

Palmately bifoliolate, trifoliolate, plurifoliolate (of several leaflets), &c., or else

Pinnately bi-, tri-, quadri-, or *pluri-foliolate* (that is, of two, three, four, five, or several leaflets), as the case may be.

FIG. 130. A twice-pinnate (abruptly) leaf of the Honey-Locust.

are *bipinnate*, i. e. *twice pinnate*, as in Fig. 130. If these leaflets were again divided in the same way, the leaf would become *thrice pinnate*, or *tripinnate*, as in many Acacias. The first divisions are called *pinnæ ;* the others, *pinnules ;* and the last, or little blades, *leaflets.*

172. So the palmate leaf, if again compounded in the same way, becomes *twice palmate*, or, as we say when the divisions are in threes, *twice ternate* (in Latin form *biternate*) ; if a third time compounded, *thrice ternate* or *triternate*. But if the division goes still further, or if the degree is variable, we simply say that the leaf is *decompound ;* either palmately or pinnately so, as the case may be. Thus, Fig. 138 represents a four times ternately compound, in other words a *ternately decompound*, leaf of our common Meadow Rue.

173. So exceedingly various are the kinds and shapes of leaves, that we have not yet exhausted the subject. We have, however, mentioned the principal terms used in describing them. Many others will be found in the glossary at the end of the volume. Some peculiar sorts of leaves remain to be noticed, which the student might not well understand without some explanation ; such as

174. **Perfoliate Leaves.** A common and simple case of this sort is found in two species of Uvularia or Bellwort, where the stem appears to run through the blade of the leaf, near one end. If we look at this plant in summer, after all the leaves are formed, we may see the meaning of this at a glance. For then we often find upon the same stem such a series of leaves as is given in Fig. 131 : the lower leaves are *perfoliate*, those next above less so ; then some (the fourth and fifth) with merely a heart-shaped clasping base, and finally one that is merely *sessile.* The leaf, we perceive, becomes perfoliate by the union of the edges of the base with each other around the stem ; just as the shield-shaped leaf, Fig.
102, comes from the union of the edges of the base of such a leaf as Fig. 101. Of the same sort are the upper leaves of most of

FIG. 131. Leaves of Uvularia (Bellwort) ; the lower ones perfoliate, the others merely clasping, or the uppermost only sessile.

the true Honeysuckles (Fig. 132): but here it is a pair of opposite leaves, with their contiguous broad bases grown together, which
makes what seems to be one round leaf, with the stem running
through its centre. This is seen to be the case, by comparing
together the upper and the lowest leaves of the same branch.
Leaves of this sort are said to be *connate-perfoliate*.

175. **Equitant Leaves.** While ordinary
leaves spread horizontally, and present
one face to the sky and the other to the
earth, there are some that present their
tip to the sky, and their faces right
and left to the horizon. Among these
are the *equitant* leaves of the Iris or
Flower-de-Luce. On careful inspection
we shall find that each leaf was formed
*folded together length-
wise*, so that what
would be the upper
surface is within, and
all grown together, ex-
cept next the bottom,
where each leaf covers
the next younger one. It was from their straddling over each other, like a man on horseback (as
is seen in the cross-section, Fig. 134), that Linnæus,
with his lively fancy, called these *equitant* leaves.

176. **Leaves with no distinction of Petiole and Blade.**
The leaves of Iris just mentioned show one form
of this. The flat but narrow
leaves of Jonquils, Daffodils,
and the like, are other instances. *Needle-shaped* leaves,
like those of the Pine (Fig.
140), Larch (Fig. 139), and
Spruce, and the *awl-shaped*
as well as the *scale-shaped*
leaves of Junipers, Red Ce-

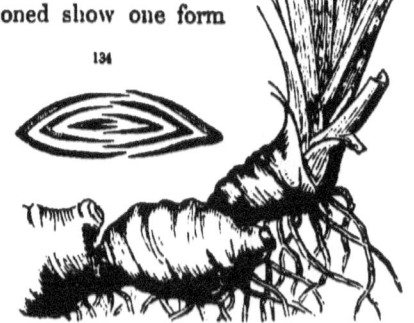

FIG. 132. Branch of a Yellow Honeysuckle, with connate-perfoliate leaves.
FIG. 133. Rootstock and equitant leaves of Iris. 134. A section across the cluster of
leaves at the bottom.

dar, and Arbor-Vitæ (Fig. 135), are different examples. These last are leaves serving for foliage, but having as little spread of surface as possible. They make up for this, however, by their immense numbers.

177. Sometimes the *petiole* expands and flattens, and takes the place of the blade; as in numerous New Holland Acacias, some of which are now common in greenhouses. Such counterfeit blades are called *phyllodia*, — meaning leaf-like bodies. They may be known from true blades by their standing edgewise, their margins being directed upwards and downwards; while in true blades the faces look upwards and downwards; excepting in equitant leaves, as already explained, and in those which are tur..ed edgewise by

135

a twist, such as those of the Callistemon or Bottle-brush Flower of our greenhouses, and other Dry Myrtles of New Holland, &c.

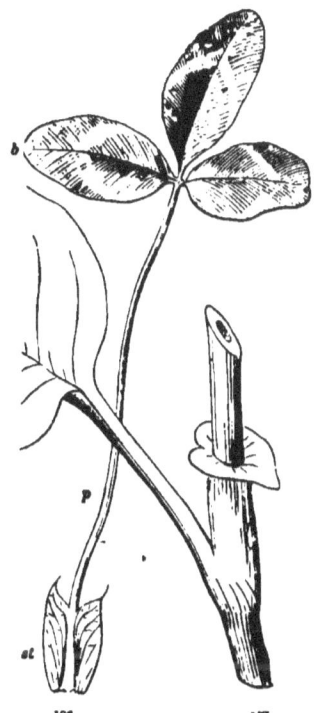

178. **Stipules,** the pair of appendages which is found at the base of the petiole in many leaves (133), should also be considered in respect to their very varied forms and appearances. More commonly they appear like little blades, on each side of the leaf-stalk, as in the Quince (Fig. 83), and more strikingly in the Hawthorn and in the Pea. Here they remain as long as the rest of the leaf, and serve for the same purpose as the blade. Very commonly they serve for bud-scales, and fall off when the leaves expand, as in the Fig-tree, and the Magnolia (where they are large and conspicuous), or soon

136   137

FIG. 135. Twig of Arbor-Vitæ, with its two sorts of leaves: viz. some awl-shaped, the others scale-like ; the latter on the branchlets, *a.*

FIG. 136. Leaf of Red Clover: *st,* stipules, adhering to the base of *p,* the petiole : *b,* blade of three leaflets.

FIG. 137. Part of stem and leaf of Prince's-Feather (Polygonum orientale) with the united sheathing stipules forming a sheath.

afterwards, as in the Tulip-tree.   In the Pea the stipules make a
very conspicuous part of the leaf; while in the Bean they are quite
small; and in the Locust they are reduced to bristles or prickles.
Sometimes the stipules are separate and distinct (Fig. 83): often
they are united with the base of the leaf-stalk, as in the Rose and
the Clover (Fig. 136): and sometimes they grow together by both
margins, so as to form a sheath around the stem, above the leaf, as
in the Buttonwood, the Dock, and almost all the plants of the
Polygonum Family (Fig. 137).

179. The sheaths of Grasses bear the blade on their summit, and
therefore represent a form of the petiole.   The small and thin appendage which is commonly found at the top of the sheath (called a
ligule) here answers to the stipule.

FIG. 138.   Ternately-decompound leaf of Meadow Rue (Thalictrum Cornuti).

LESSON X.

THE ARRANGEMENT OF LEAVES.

180. UNDER this head we may consider,— 1. the arrangement of leaves on the stem, or what is sometimes called PHYLLOTAXY (from two Greek words meaning *leaf-order*) ; and 2. the ways in which they are packed together in the bud, or their VERNATION (the word meaning their spring state).

181. **Phyllotaxy.** As already explained (48, 49), leaves are arranged on the stem in two principal ways. They are either

Alternate (Fig. 131, 143), that is, one after another, only a single leaf arising from each node or joint of the stem ; or

Opposite (Fig. 147), when there is a pair of leaves on each joint of the stem ; one of the two leaves being in this case always situated exactly on the opposite side of the stem from the other. A third, but uncommon arrangement, may be added ; namely, the

Whorled, or *verticillate* (Fig. 148), when there are three or more leaves in a circle (*whorl* or *verticil*) on one joint of stem. But this is only a variation of the opposite mode; or rather the latter arrangement is the same as the whorled, with the number of the leaves reduced to two in each whorl.

182. Only one leaf is ever produced from the same point. When two are borne on the same joint, they are always on opposite sides of the stem, that is, are separated by half the circumference ; when in whorls of three, four, five, or any other number, they are equally distributed around the joint of stem, at a distance of one third, one fourth, or one fifth of the circumference from each other, according to their number. So they always have the greatest possible divergence from each other. Two or more leaves be-longing to the same joint of stem never stand side by side, or one above the other, in a cluster.

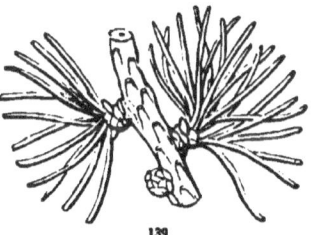

183. What are called *clustered* or *fascicled* leaves, and which

FIG. 139. Clustered or fascicled leaves of the Larch

appear to be so, are always the leaves of a whole branch which remains so very short that they are all crowded together in a bundle or rosette; as in the spring leaves of the Barberry and of the Larch (Fig. 139). In these cases an examination shows them to be nothing else than alternate leaves, very much crowded on a short spur; and some of these spurs are seen in the course of the season to lengthen into ordinary shoots with scattered alternate leaves. So, likewise, each cluster of two or three needle-shaped leaves in Pitch Pines (as in Fig. 140), or of five leaves in White Pine, answers to a similar, extremely short branch, springing from the axil of a thin and slender scale, which represents a leaf of the main shoot. For Pines produce two kinds of leaves; — 1. primary, the proper leaves of the shoots, not as foliage, but in the shape of delicate scales in spring, which soon fall away; and 2. secondary, the *fascicled* leaves, from buds in the axils of the former, and these form the actual foliage.

184. **Spiral Arrangement of Leaves.** If we examine any alternate-leaved stem, we shall find that the leaves are placed upon it in symmetrical order, and in a way perfectly uniform for each species, but different in different plants. If we draw a line from the *insertion* (i. e. the point of attachment) of one leaf to that of the next, and so on, this line will wind spirally around the stem as it rises, and in the same species will always have just the same number of leaves upon it for each turn round the stem. That is, any two successive leaves will always be separated from each other by just an equal portion of the circumference of the stem. The distance in *height* between any two leaves may vary greatly, even on the same shoot, for that depends upon the length of the *internodes* or spaces between each leaf; but the distance as measured around the circumference (in other words, the *angular divergence*, or angle formed by any two successive leaves) is uniformly the same.

185. The greatest possible divergence is, of course, where the second leaf stands on exactly the opposite side of the stem from the first, the third on the side opposite the second, and therefore over the

first, and the fourth over the second. This brings all the leaves into two ranks, one on one side of the stem and one on the other; and is therefore called the *two-ranked* arrangement. It occurs in all Grasses, — in Indian Corn, for instance; also in the Spiderwort, the Bellwort (Fig. 131) and Iris (Fig. 132), in the Basswood or Lime-tree, &c. This is the simplest of all arrangements.

186. Next to this is the *three-ranked* arrangement, such as we see in Sedges, and in the Veratrum or White Hellebore. The plan of it is shown on a Sedge in Fig. 141, and in a diagram or cross-section underneath, in Fig. 142. Here the second leaf is placed one third of the way round the stem, the third leaf two thirds of the way round, the fourth leaf accordingly directly over the first, the fifth over the second, and so on. That is, three leaves occur in each turn round the stem, and they are separated from each other by one third of the circumference.

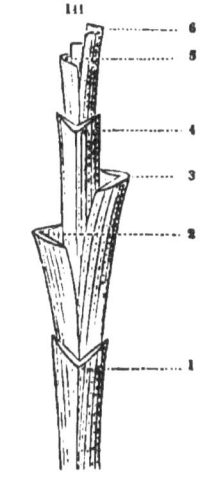

187. The next and one of the most common is the *five-ranked* arrangement; which is seen in the Apple (Fig. 143), Cherry, Poplar, and the greater part of our trees and shrubs. In this case the line traced from leaf to leaf will pass twice round the stem before it reaches a leaf situated directly over any below (Fig. 144). Here the sixth leaf is over the first; the leaves stand in five perpendicular ranks, equally distant from each other; and the distance between any two successive leaves is just two fifths of the circumference of the stem.

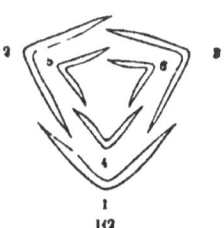

188. The five-ranked arrangement is expressed by the fraction ⅖. This fraction denotes the divergence of the successive leaves, i. e. the angle they form with each other: the numerator also expresses the number of turns made round the stem by the spiral line in completing one cycle or set of leaves, namely 2; and the denominator gives the number of leaves in each cycle, or the number of perpendicular

ranks, namely 5. In the same way the fraction $\frac{1}{2}$ stands for the two-ranked mode, and $\frac{1}{3}$ for the three-ranked: and so these different sorts are expressed by the series of fractions $\frac{1}{2}$, $\frac{1}{3}$, $\frac{2}{5}$. And the other cases known follow in the same numerical progression.

189. The next is the *eight-ranked* arrangement, where the ninth leaf stands over the first, and three turns are made around the stem to reach it; so it is expressed by the fraction $\frac{3}{8}$. This is seen in the Holly, and in the common Plantain. Then comes the *thirteen-ranked* arrangement, in which the fourteenth leaf is over the first, after five turns around the stem. Of this we have a good example in the common Houseleek (Fig. 146).

190. The series so far, then, is $\frac{1}{2}$, $\frac{1}{3}$, $\frac{2}{5}$, $\frac{3}{8}$, $\frac{5}{13}$; the numerator and the denominator of each fraction being those of the two next preceding ones added together. At this rate the next higher should be $\frac{8}{21}$, then $\frac{13}{34}$, and so on; and in fact just such cases are met with, and (commonly) no others. These higher sorts are found in the Pine Family, both in the leaves and the cones (Fig. 324), and in many other plants with small and crowded leaves. But the number of the ranks, or of leaves in each cycle, can here rarely be made out by direct inspection: they may be ascertained, however, by certain simple mathematical computations, which are rather too technical for these Lessons.

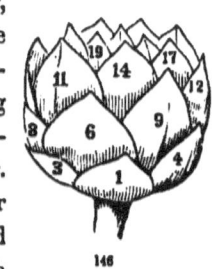

FIG. 143. Shoot with its leaves 5-ranked, the sixth leaf over the first; as in the Apple-tree.

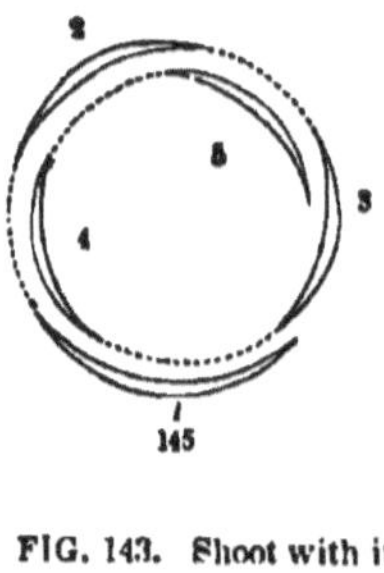

FIG. 144. Diagram of this arrangement, with a spiral line drawn from the attachment of one leaf to the next, and so on; the parts on the side turned from the eye are fainter.

FIG. 145. A ground-plan of the same; the section of the leaves similarly numbered; a dotted line drawn from the edge of one leaf to that of the next completes the spiral.

FIG. 146. A young plant of the Houseleek, with the leaves (not yet expanded) numbered, and exhibiting the 13-ranked arrangement

191. The arrangement of opposite leaves (181) is usually very simple. The second pair is placed over the intervals of the first; the third over the intervals of the second, and so on (Fig. 147); the successive pairs thus crossing each other, — commonly at right angles, so as to make four upright rows. And *whorled* leaves (Fig. 148) follow a similar plan.

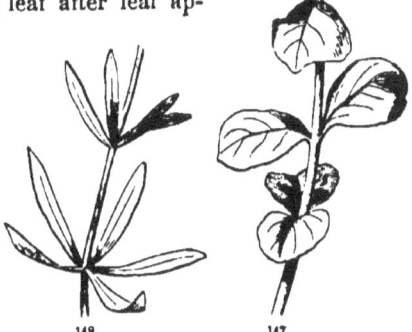

192. So the place of every leaf on every plant is fixed beforehand by unerring mathematical rule. As the stem grows on, leaf after leaf appears exactly in its predestined place, producing a perfect symmetry ; — a symmetry which manifests itself not in one single monotonous pattern for all plants, but in a definite number of forms exhibited by different species, and arithmetically expressed by the series of fractions, $\frac{1}{2}$, $\frac{1}{3}$, $\frac{2}{5}$, $\frac{3}{8}$, $\frac{5}{13}$, $\frac{8}{21}$, &c., according as the formative energy in its spiral course up the developing stem lays down at corresponding intervals 2, 3, 5, 8, 13, or 21 ranks of alternate leaves.

148    147

193. **Vernation**, sometimes called *Præfoliation*, relates to the way in which leaves are disposed in the bud (180). It comprises two things ; — 1st, the way in which each separate leaf is folded, coiled, or packed up in the bud ; and 2d, the arrangement of the leaves in the bud with respect to one another. The latter of course depends very much upon the phyllotaxy, i. e. the position and order of the leaves upon the stem. The same terms are used for it as for the arrangement of the leaves of the flower in the flower-bud: so we may pass them by until we come to treat of the flower in this respect.

194. As to each leaf separately, it is sometimes *straight* and open in vernation, but more commonly it is either *bent, folded,* or *rolled up.* When the upper part is bent down upon the lower, as the young blade in the Tulip-tree is bent upon the leafstalk; it is said to be *inflexed* or *reclined* in vernation. When folded

FIG. 147. Opposite leaves of the Spindle-tree or Burning-bush.
FIG. 148. Whorled or verticillate leaves of Galium or Bedstraw.

by the midrib so that the two halves are placed face to face, it is *conduplicate* (Fig. 149), as in the Magnolia, the Cherry, and the Oak: when folded back and forth like the plaits of a fan, it is *plicate* or *plaited* (Fig. 150), as in the Maple and Currant.   If rolled, it may be so either from the tip downwards, as in Ferns and the Sundew (Fig. 154), when in unrolling it resembles the head of a crosier, and is said to be *circinate;* or it may be rolled up parallel with the axis, either from one edge into a coil, when it is *convolute* (Fig. 151), as in the Apricot and Plum, or rolled f.om both edges towards the midrib; — sometimes inwards, when it is *involute* (Fig. 152), as in the Violet and Water-Lily; sometimes outwards, when it is *revolute* (Fig. 153),.in the Rosemary and Azalea.   The figures are diagrams, representing sections through the leaf, in the way they were represented by Linnæus.

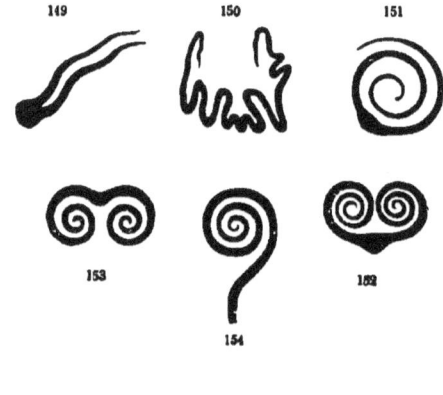

149   150   151

153   152

154

LESSON XI.

THE ARRANGEMENT OF FLOWERS ON THE STEM, OR INFLO-
RESCENCE.

195. Thus far we have been considering the *vegetation* of the plant, and studying those parts, viz. root, stem, and leaves, by which it increases in size and extent, and serves the purpose of its individual life.   But after a time each plant produces a different set of organs, — viz. flowers, fruit, and seed, — subservient to a different purpose, that is, the increase in numbers, or the continuance of the

species. The plant reproduces itself in new individuals by seed. Therefore the *seed*, and the *fruit* in which the seed is formed, and the *flower*, from which the fruit results, are named the *Organs of Reproduction* or *Fructification*. These we may examine in succession. We begin, of course, with the flower. And the first thing to consider is the

196. **Inflorescence**, or the mode of flowering, that is, the situation and arrangement of blossoms on the plant. Various as this arrangement may seem to be, all is governed by a simple law, which is easily understood. As the position of every leaf is fixed beforehand by a mathematical law which prescribes where it shall stand (192), so is that of every blossom; — and by the same law in both cases. For flowers are buds, developed in a particular way; and flower-buds occupy the position of leaf-buds, and no other   As leaf-buds are either terminal (at the summit of a stem or branch, 42), or axillary (in the axil of a leaf, 43), so likewise

197. Flowers are either *terminal* or *axillary*. In blossoming as in vegetation we have only buds terminating (i. e. on the summit of) stems or branches, and buds from the axils of leaves. But while the same plant commonly produces both kinds of leaf-buds, it rarely bears flowers in both situations. These are usually either all axillary or all terminal; — giving rise to two classes of inflorescence, viz. the *determinate* and the *indeterminate*.

198. **Indeterminate Inflorescence** is that where the flowers all arise from axillary buds; as in Fig. 155, 156, 157, &c.; and the reason why it is called indeterminate (or *indefinite*) is, that while the axillary buds give rise to flowers, the terminal bud goes on to grow, and continues the stem indefinitely.

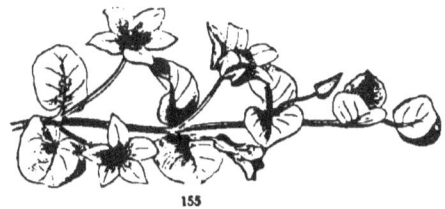

155

199. Where the flowers arise, as in Fig. 155, singly from the axils of the ordinary leaves of the plant, they do not form flower-clusters, but are *axillary* and *solitary*. But when several or many flowers are produced near each other, the accompanying leaves are usually of smaller size, and often of a different shape or character: then they are called *bracts*; and the flowers thus brought together

FIG. 155   Moneywort (Lysimachia nummularia) of the gardens, with axillary flowers-

7 *

form one cluster or inflorescence. The sorts of inflorescence of the indeterminate class which have received separate names are chiefly the following: viz. the *Raceme*, the *Corymb*, the *Umbel*, the *Spike*, the *Head*, the *Spadix*, the *Catkin*, and the *Panicle*.

200. Before illustrating these, one or two terms, of common occurrence, may be defined. A flower (or other body) which has no stalk to support it, but which sits directly on the stem or axis it proceeds from, is said to be *sessile*. If it has a stalk, this is called its *peduncle*. If the whole flower-cluster is raised on a stalk, this is called the peduncle, or the *common peduncle* (Fig. 156, *p*); and the

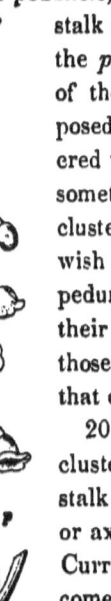

stalk of each particular flower, if it have any, is called the *pedicel* or *partial peduncle* (*p'*). The portion of the general stalk along which flowers are disposed is called the *axis of inflorescence*, or, when covered with sessile flowers, the *rhachis* (back-bone), and sometimes the *receptacle*. The leaves of a flower-cluster generally are termed *bracts*. But when we wish particularly to distinguish them, those on the peduncle, or main axis, and which have a flower in their axil, take the name of *bracts* (Fig. 156, *b*); and those on the pedicels or partial flower-stalks, if any, that of *bractlets* (Fig. 156, *b'*).

201. A Raceme (Fig. 156, 157) is that form of flower-cluster in which the flowers, each on their own foot-stalk or pedicel, are arranged along a common stalk or axis of inflorescence; as in the Lily of the Valley, Currant, Choke-Cherry, Barberry, &c. Each flower comes from the axil of a small leaf, or bract, which, however, is often so small that it might escape notice, and which sometimes (as in the Mustard Family) disappears altogether. The lowest blossoms of a raceme are of course the oldest, and therefore open first, and the order of blossoming is *ascending*, from the bottom to the top. The summit, never being stopped by a terminal flower, may go on to grow, and often does so (as in the common Shepherd's Purse), producing lateral flowers one after another the whole summer long.

202. All the various kinds of flower-clusters pass one into another

---

FIG. 156. A Raceme, with a general peduncle (*p*), pedicels (*p'*), bracts (*b*), and bractlets (*b'*).

by intermediate gradations of every sort.   For instance, if we lengthen the lower pedicels of a raceme, and keep the main axis rather short, it is converted into

203. A Corymb (Fig. 158).   This is the same as a raceme, except that it is flat and broad, either convex, or level-topped, as in the Hawthorn, owing to the lengthening of the lower pedicels while the uppermost remain shorter.

204. The main axis of a corymb is short, at least in comparison with the lower pedicels.   Only suppose it to be so much contracted that the bracts are all brought into a cluster or circle, and the corymb becomes

205. An Umbel (Fig. 159), — as in the Milkweed and Primrose, — a sort of flower-cluster where the pedicels all spring apparently from the same point, from the top of the peduncle, so as to resemble, when spreading, the rays of an umbrella, whence the name.   Here the pedicels are sometimes called the *rays* of the umbel.   And the bracts, when brought in this way into a cluster or circle, form what is called an *involucre*.

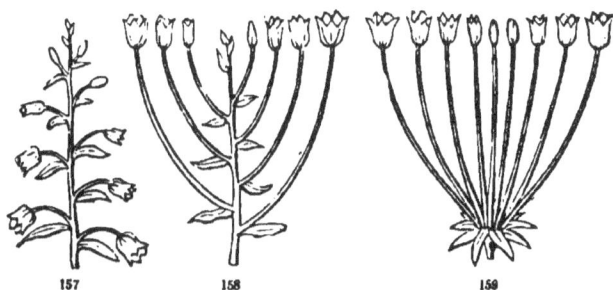

157   158   159

206. For the same reason that the order of blossoming in a raceme is ascending (201), in the corymb and umbel it is *centripetal*, that is, it proceeds from the margin or circumference regularly towards the centre; the lower flowers of the former answering to the outer ones of the latter.   Indeterminate inflorescence, therefore, is said to be centripetal in evolution.   And by having this order of blossoming, all the sorts may be distinguished from those of the other, or the determinate class.   In all the foregoing cases the flowers are raised on pedicels.   These, however, are very short in many instances, or are wanting altogether; when the flowers are *sessile* (200).   They are so in

FIG. 157.   A raceme.   158.   A corymb.   159.   An umbel.

207. **The Spike.** This is a flower-cluster with a more or less lengthened axis, along which the flowers are sessile or nearly so; as in the Mullein and the Plantain (Fig. 160). It is just the same as a raceme, therefore, without any pedicels to the flowers.

208. **The Head** is a round or roundish cluster of flowers which are sessile on a very short axis or receptacle, as in the Button-ball, Button-bush (Fig. 161), and Red Clover. It is just what a spike would become if its axis were shortened; or an umbel, if its pedicels were all shortened until the flowers became sessile or apparently so. The head of the Button-bush (Fig. 161) is naked; but that of the Thistle, of the Dandelion, the Cichory (Fig. 221), and the like, is surrounded by empty bracts, which form an *involucre*. Two particular forms of the spike and the head have received particular names, namely, the *Spadix* and the *Catkin*.

209. **A Spadix** is nothing but a fleshy spike or head, with small and often imperfect flowers, as in the Calla, the Indian Turnip

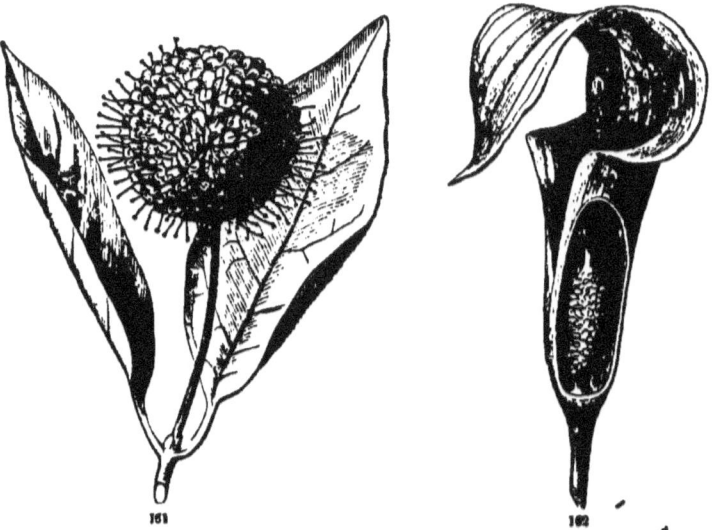

(Fig. 162), Sweet Flag, &c. It is commonly covered by a peculiar enveloping leaf, called a *spathe*.

FIG. 160. Spike of the common Plantain or Ribwort.
FIG. 161. Head of the Button-bush (Cephalanthus).
FIG. 162. Spadix and spathe of the Indian Turnip; the latter cut through below.

210. **A Catkin or Ament** is the name given to the scaly sort of spike of the Birch and Alder, the Willow and Poplar, and one sort of flower-clusters of the Oak, Hickory, and the like; — on which account these are called *Amentaceous* trees.

211. Sometimes these forms of flower-clusters become *compound.* For example, the stalks which, in the simple umbel such as has been described (Fig. 159), are the pedicels of single flowers, may themselves branch in the same way at the top, and so each become the support of a smaller umbel; as is the case in the Parsnip, Caraway, and almost the whole of the great family of what are called *Umbelliferous* (i. e. umbel-bearing) plants. Here the whole is termed a *compound umbel;* and the smaller or *partial* umbels take the name in English of *umbellets.* The *general involucre*, at the base of the main umbel, keeps that name; while that at the base of each umbellet is termed a *partial involucre* or an *involucel.*

212. So a corymb (Fig. 158) with its separate stalks branching again, and bearing smaller clusters of the same sort, is a *compound corymb;* of which the Mountain Ash is a good example. A raceme where what would be the pedicels of single flowers become stalks, along which flowers are disposed on their own pedicels, forms a *compound raceme,* as in the Goat's-beard and the False Spikenard. But when what would have been a raceme or a corymb branches irregularly into an open and more or less compound flower-cluster, we have what is called

213. **A Panicle** (Fig. 163); as in the Oat and in most common Grasses. Such a raceme as that of the diagram, Fig. 156, would be changed into a panicle like Fig. 163, by the production of a flower from the axil of each of the bractlets *b*.

214. **A Thyrsus** is a compact panicle of a pyramidal or oblong shape; such as a bunch of grapes, or the cluster of the Lilac or Horsechestnut.

215. **Determinate Inflorescence** is that in which the flowers are from terminal buds. The simplest case is where a stem bears a solitary, terminal flower, as in Fig. 163ᵃ. This stops the growth of

FIG. 163. A Panicle

S & F—5

the stem; for its terminal bud, being changed into a blossom, can no more lengthen in the manner of a leaf-bud.   Any further growth

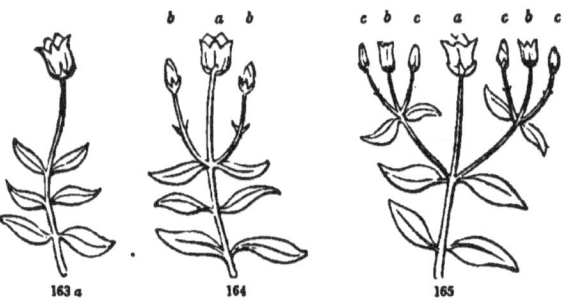

must be from axillary buds developing into branches.   If such branches are leafy shoots, at length terminated by single blossoms, the inflorescence still consists of solitary flowers at the summit of the stem and branches.   But if the flowering branches bear only bracts in place of ordinary leaves, the result is the kind of flower-cluster called

216. **A Cyme.**   This is commonly a flat-topped or convex flower-cluster, like a corymb, only the blossoms are from terminal buds.   Fig. 164 illustrates the simplest cyme in a plant with opposite leaves, namely, with three flowers.   The middle flower, *a*, terminates the stem; the two others, *b b*, terminate short branches, one from the axil of each of the uppermost leaves; and being later than the middle one, the flowering proceeds from the centre outwards, or is *centrifugal;* — just the opposite of the indeterminate mode, or that where all the flower-buds are axillary.   If flowering branches appear from the axils below, the lower ones are the later, so that the order of blossoming continues *centrifugal* or *descending* (which is the same thing), as in Fig. 166, making a sort of reversed raceme; — a kind of cluster which is to the true raceme just what the flat cyme is to the corymb.

217. Wherever there are bracts or leaves, buds may be produced from their axils and appear as flowers.   Fig. 165 represents the case where the branches, *b b,* of Fig. 164, each with a pair of small

FIG. 163 *a.* Diagram of an opposite-leaved plant, with a single terminal flower. 164 Same, with a cyme of three flowers ; *a*, the first flower, of the main axis ; *b b*, those of branches. 165. Same, with flowers of the third order, *c c.* 166. Same, with flowers only of the second order from all the axils ; the central or uppermost opening first, and so on downwards.

leaves or bracts about their middle, have branched again, and produced the branchlets and flowers *c c*, on each side. It is the continued repetition of this which forms the full or compound cyme, such as that of the Laurustinus, Hobblebush, Dogwood, and Hydrangea (Fig. 167).

218. A **Fascicle**, like that of the Sweet-William and Lychnis of the gardens, is only a cyme with the flowers much crowded, as it were, into a bundle.

219. A **Glomerule** is a cyme still more compacted, so as to form a sort of head. It may be known from a true head by the flowers not expanding centripetally, that is, not from the circumference towards the centre, or from the bottom to the top.

220. The illustrations of determinate or *cymose* inflorescence have been taken from plants with opposite leaves, which give rise to the most regular cymes. But the Rose, Cinquefoil, Buttercup, and the like, with alternate leaves, furnish equally good examples of this class of flower-clusters.

221. It may be useful to the student to exhibit the principal sorts of inflorescence in one view, in the manner of the following

Analysis of Flower-Clusters.

I. INDETERMINATE OR CENTRIPETAL. (198.)
  Simple; and with the
  Flowers borne on pedicels,

Along the sides of a lengthened axis,	RACEME,	201
Along a short axis ; lower pedicels lengthened,	CORYMB,	203.
Clustered on an extremely short axis,	UMBEL,	205.

  Flowers sessile, without pedicels (206),

Along an elongated axis,	SPIKE,	207.
On a very short axis,	HEAD,	208.
with their varieties, the SPADIX, 209, and	CATKIN,	210.
Branching irregularly,	PANICLE,	213.
with its variety, the	THYRSUS,	214.

II. DETERMINATE OR CENTRIFUGAL. (215.)

Open, mostly flat-topped or convex,	CYME,	216.
Contracted into a bundle,	FASCICLE,	218.
Contracted into a sort of head,	GLOMERULE,	219.

222. The numbers refer to the paragraphs of this Lesson. The various sorts run together by endless gradations in different plants. The botanist merely designates the leading kinds by particular names. Even the two classes of inflorescence are often found combined in the same plant. For instance, in the whole Mint Family,

the flower-clusters are centrifugal, that is, are cymes or fascicles ; but they are themselves commonly disposed in spikes or racemes, which are centripetal, or develop in succession from below upwards.

LESSON XII.

THE FLOWER: ITS PARTS OR ORGANS.

223. HAVING considered, in the last Lesson, the arrangement of flowers on the stem, or the places from which they arise, we now direct our attention to the flower itself.

224. **Nature and Use of the Flower.** The object of the flower is the production of seed. The flower consists of all those parts, or *organs,* which are subservient to this end. Some of these parts are necessary to the production of seed. Others serve merely to protect or support the more essential parts.

FIG. 167. Cyme of the Wild Hydrangea (with neutral flowers in the border).

225. **The Organs of the Flower** are therefore of two kinds; namely, first, the *protecting organs*, or *leaves of the flower*, — also called the *floral envelopes*, — and, second, the *essential organs*. The latter are situated within or a little above the former, and are enclosed by them in the bud.

226. **The Floral Envelopes** in a complete flower are double ; that is, they consist of two whorls (181), or circles of leaves, one above or within the other. The outer set forms the *Calyx;* this more commonly consists of green or greenish leaves, but not always. The inner set, usually of a delicate texture, and of some other color than green, and in most cases forming the most showy part of the blossom, is the *Corolla.*

227. The floral envelopes, taken together, are sometimes called the *Perianth.* This name is not much used, however, except in cases where they form only one set, at least in appearance, as in the Lily, or where, for some other reason, the limits between the calyx and the corolla are not easily made out.

228. Each leaf or separate piece of the corolla is called a *Petal ;* each leaf of the calyx is called a *Sepal.* The sepals and the petals — or, in other words, the leaves of the blossom — serve to protect, support, or nourish the parts within. They do not themselves make a perfect flower.

229. Some plants, however, naturally produce, besides their perfect flowers, others which consist only of calyx and corolla (one or both), that is, of leaves. These, destitute as they are of the essential organs, and incapable of producing seed, are called *neutral* flowers. We have an example in the flowers round the margin of the cyme of the Hydrangea (Fig. 167), and of the Cranberry-Tree, or Snowball, in their wild state. By long cultivation in gardens the whole cluster has been changed into showy, but useless, neutral flowers, in these and some other cases. What are called *double flowers*, such as full Roses (Fig. 173), Buttercups, and Camellias, are blossoms which, under the gardener's care, have developed with all their essential organs changed into petals. But such flowers are always in an unnatural or monstrous condition, and are incapable of maturing seed, for want of

230. **The Essential Organs.** These are likewise of two kinds, placed one above or within the other ; namely, first, the *Stamens* or fertilizing organs, and, second, the *Pistils*, which are to be fertilized and bear the seeds.

8

231. Taking them in succession, therefore, beginning from below, or at the outside, we have (Fig. 168, 169), first, the *calyx* or outer

circle of leaves, which are individually termed *sepals* (*a*); secondly, the *corolla* or inner circle of delicate leaves, called *petals* (*b*); then a set of *stamens* (*c*); and in the centre one or more pistils (*d*). The end of the flower-stalk, or the short axis, upon which all these parts stand, is called the *Torus* or *Receptacle*.

232. We use here for illus-

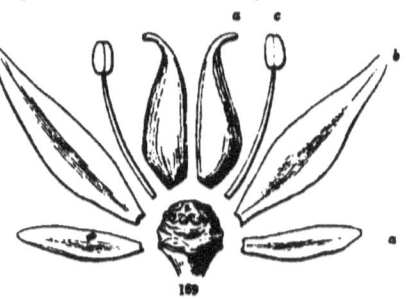

tration the flower of a species of Stonecrop (Sedum ternatum),—which is a common plant wild in the Middle States, and in gardens almost everywhere, — because, although small, it exhibits all the parts in a perfectly simple and separate state, and so answers for a sort of pattern flower, better than any larger one that is common and well known.

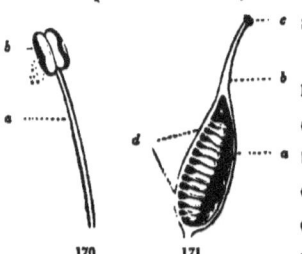

233. A **Stamen** consists of two parts, namely, the *Filament* or stalk (Fig. 170, *a*), and the *Anther* (*b*). The latter is the only essential part. It is a case, commonly with two lobes or cells, each opening lengthwise by a slit, at the proper time, and discharging a powder or dust-like substance, usually of a yellow color. This powder is the *Pollen*, or fertilizing matter, to produce which is the sole office of the stamen.

234. A **Pistil** is distinguished into three parts; namely,—beginning from below,—the *Ovary*, the *Style*, and the *Stigma*. The *Ovary* is the hollow case or young pod (Fig. 171, *a*), containing rudimentary seeds, called *Ovules* (*d*). Fig. 172, representing a pistil like that of

FIG. 168. Flower of a Stonecrop : Sedum ternatum.
FIG. 169. Two parts of each kind of the same flower, displayed and enlarged.
FIG. 170. A stamen : *a*, the filament ; *b*, the anther, discharging pollen.
FIG. 171. A pistil divided lengthwise, showing the interior of the ovary, *a*, and its ovules, *d* ; *b*, the style ; *c*, stigma.
FIG. 172. A pistil, enlarged : the ovary cut across to show the ovules within.
FIG. 173. "Double" Rose ; the essential organs all replaced by petals.

Fig. 169, *d*, but on a larger scale, and with the ovary cut across, shows the ovules as they appear in a transverse section. The *style* (Fig. 171, *b*) is the tapering part above, sometimes long and slender, sometimes short, and not rarely altogether wanting, for it is not an essential part, like the two others. The *stigma* (*c*) is the tip or some other portion of the style (or of the top of the ovary when there is no distinct style), consisting of loose tissue, not covered, like the rest of the plant, by a skin or epidermis. It is upon the stigma that the pollen falls; and the result is, that the ovules contained in the ovary are fertilized and become *seeds*, by having an embryo (16) formed in them. To the pistil, therefore, all the other organs of the blossom are in some way or other subservient: the stamens furnish pollen to fertilize its ovules; the corolla and the calyx form coverings which protect the whole.

172

234ᵃ. These are all the parts which belong to any flower. But these parts appear under a variety of forms and combinations, some of them greatly disguising their natural appearance. To understand the flower, therefore, under whatever guise it may assume, we must study its plan.

173

LESSON XIII.

THE PLAN OF THE FLOWER.

235. THE FLOWER, like every other part of the plant, is formed upon a *plan*, which is essentially the same in all blossoms; and the student should early get a clear idea of the plan of the flower. Then the almost endless varieties which different blossoms present will be at once understood whenever they occur, and will be regarded with a higher interest than their most beautiful forms and richest colors are able to inspire.

236. We have already become familiar with the plan of the vegetation; — with the stem, consisting of joint raised upon joint, each bearing a leaf or a pair of leaves; with the leaves arranged in symmetrical order, every leaf governed by a simple arithmetical law, which fixes beforehand the precise place it is to occupy on the stem; and we have lately learned (in Lesson 11) how the position of each blossom is determined beforehand by that of the leaves; so that the shape of every flower-cluster in a bouquet is given by the same simple mathematical law which arranges the foliage. Let us now contemplate the flower in a similar way. Having just learned what parts it consists of, let us consider the plan upon which it is made, and endeavor to trace this plan through some of the various forms which blossoms exhibit to our view.

237. In order to give at the outset a correct idea of the blossom, we took, in the last Lesson, for the purpose of explaining its parts, a *perfect, complete, regular*, and *symmetrical* flower, and one nearly as *simple* as such a flower could well be. Such a blossom the botanist regards as

238. A Typical Flower, that is, a *pattern flower*, because it well exemplifies the plan upon which all flowers are made, and serves as what is called a *type*, or standard of comparison.

239. Another equally good typical flower (except in a single respect, which will hereafter be mentioned), and one readily to be obtained in the summer, is that of the Flax (Fig. 174). The parts differ in shape from those of the Stonecrop; but the whole plan is evidently just the same in both. Only, while the Stonecrop has ten stamens, or in many flowers eight stamens, — in all cases just twice

as many as there are petals, — the Flax has only five stamens, or just as many as the petals.   Such flowers as these are said to be

Perfect, because they are provided with both kinds of essential organs (230), namely, stamens and pistils ;

Complete, because they have all the *sorts* of organs which any flower has, namely, both calyx and corolla, as well as stamens and pistils ;

Regular, because all the parts of each set are alike in shape and size ; and

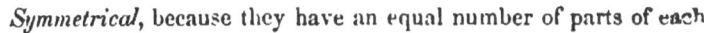

Symmetrical, because they have an equal number of parts of each sort, or in each set or circle of organs.   That is, there are five sepals, five petals, five stamens, or in the Stonecrop ten stamens (namely, two sets of five each), and five pistils.

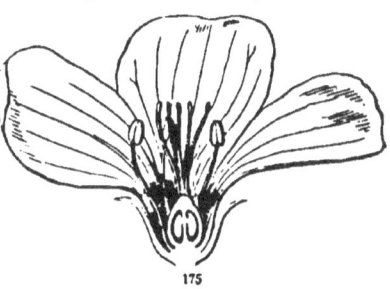

240. On the other hand, many flowers do not present this perfect symmetry and reg-ularity, or this completeness of parts.   Accord-ingly, we may have

241. **Imperfect, or Separated Flowers;** which are those where the stamens and pistils are in separate blossoms; that is, one sort of flowers has stamens and no pistils, and another has pistils and no sta-mens, or only imperfect ones.   The blossom which has stamens but no pistils is called a *staminate* or *sterile* flower (Fig. 176) ; and the corresponding one with pistils but no stamens is called a *pistillate* or *fertile* flower (Fig. 177).   The two sorts may grow on distinct plants, from different roots, as they do in the Willow and Poplar, the Hemp, and the Moonseed

FIG. 174. Flowers of the common Flax: a perfect, complete, regular, and symmetrical blossom, all its parts in fives. 175. Half of a Flax-flower divided lengthwise, and enlarged. FIG. 176. Staminate flower of Moonseed (Menispermum Canadense). 177. Pistillate flower of the same.

8 *

(Fig. 176, 177); when the flowers are said to be *diœcious* (from two Greek words meaning in two households). Or the two may occur

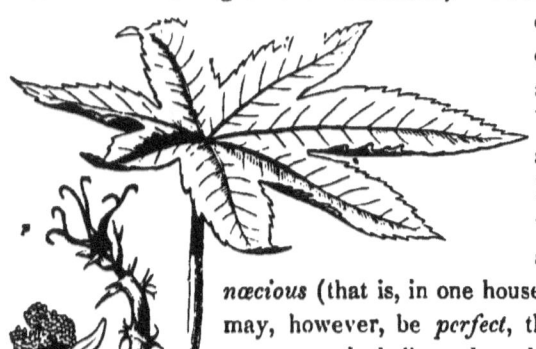

on the same plant or the same stem, as in the Oak, Walnut, Nettle, and the Castor-oil Plant (Fig. 178); when the flowers are said to be *monœcious* (that is, in one household). A flower may, however, be *perfect*, that is, have both stamens and pistils, and yet be *incomplete*.

242. **Incomplete Flowers** are those in which one or both sorts of the floral envelopes, or leaves of the blossom, are wanting. Sometimes only one sort is wanting, as in the Castor-oil Plant (Fig. 178) and in the Anemone (Fig. 179). In this case the missing sort is always supposed to be the inner, that is, the corolla; and accordingly such flowers are said to be *apetalous* (meaning without petals). Occasionally both the corolla and the calyx are wanting, when the flower has no proper coverings or floral envelopes at all. It is then said to be *naked*, as in the Lizard's-tail (Fig. 180), and in the Willow.

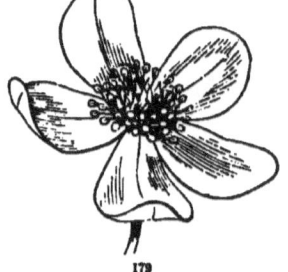

243. Our two pattern flowers (Fig. 168, 174) are regular and symmetrical (239). We commonly expect this to be the case in living things. The corresponding parts of plants, like the limbs or members of animals, are generally alike, and the whole arrangement is symmetrical. This symmetry pervades the blossom, especially. But the student may often fail to perceive

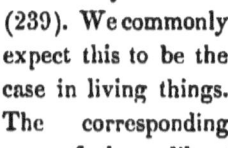

it, at first view, at least in cases where the plan is more or less obscured by the leaving out (*obliteration*) of one or more of the members of the same set, or by some inequality in their size and shape. The latter circumstance gives rise to

244. **Irregular Flowers.** This name is given to blossoms in which the different members of the same sort, as, for example, the petals or the stamens, are unlike in size or in form. We have familiar

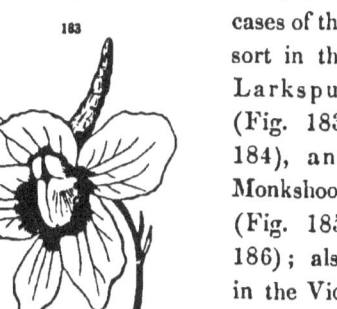

cases of the sort in the Larkspur (Fig. 183, 184), and Monkshood (Fig. 185, 186); also in the Violet (Fig. 181, 182). In the latter it is the corolla principally which is irregular, one of the petals being larger than the rest, and extended at the base into a hollow protuberance or spur. In the Larkspur (Fig. 183), both the calyx and the corolla partake of the irregularity. This and the Monkshood are likewise good examples of

245. **Unsymmetrical Flowers.** We call them unsymmetrical, when the different sets of organs do not agree in the number of their parts. The irregular calyx of Larkspur (Fig. 183, 184) consists of five sepals, one of which, larger than the rest, is prolonged behind into a large spur; but the corolla is made of only four petals (of two shapes);

FIG. 181.   Flower of a Violet.   182. Its calyx and corolla displayed: the five smaller parts are the sepals; the five intervening larger ones are the petals.

FIG. 183.   Flower of a Larkspur.   184. Its calyx and corolla displayed; the five larger pieces are the sepals; the four smaller, the petals.

the fifth, needed to complete the symmetry, being left out. And the Monkshood (Fig. 185, 186) has five very dissimilar sepals,

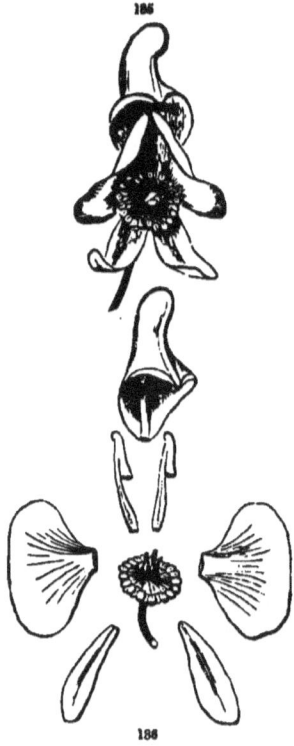

and a corolla of only two, very small, curiously-shaped petals; the three need-ed to make up the symmetry being left out. For a flower which is unsymmet-rical but regular, we may take the com-mon Purslane, which has a calyx of only two sepals, but a corolla of five petals, from seven to twelve stamens, and about six styles. The Mustard, and all flowers of that family, are un-symmetrical as to the stamens, these being six in number (Fig. 188, while the leaves of the blossom (sepals and petals) are each only four (Fig. 187). Here the stamens are *irregular* also, two of them being shorter than the other four.

246. Numerical Plan of the Flower. Although not easy to make out in all cases, yet generally it is plain to see that each blossom is based upon a particular number, which runs through all or most of its parts. And a prin-cipal thing which a botanist notices when examin-ing a flower is its numerical plan. It is upon this that the symmetry of the blossom depends. Our two pattern flowers, the Stonecrop (Fig. 168) and the Flax (Fig. 174), are based upon the number five, which is exhibited in all their parts. Some flowers of this same Stonecrop have their parts in fours, and then that number runs throughout; namely, there are four sepals, four petals, eight stamens (two sets), and four pistils. The Mustard (Fig. 187, 188), Radish,

FIG. 185. Flower of a Monkshood. 186. Its parts displayed : the five larger pieces are the sepals ; the two small ones under the hood are petals ; the stamens and pistils are in the tentre.
FIG. 187. Flower of Mustard. 188. Its stamens and pistil separate and enlarged.

&c., also have their flowers constructed on the plan of four as to the calyx and corolla, but this number is interfered with in the stamens, either by the leaving out of two sta-
mens (which would complete two sets), or in some other way. Next to five, the most common number in flowers is three. On this number the flowers of Lily, Crocus, Iris, Spiderwort, and Trillium (Fig. 189) are constructed. In the Lily and Crocus the leaves of the flower at first view appear to be six in one set; but the bud or just-

189

opening blossom plainly shows these to consist of an outer and an inner circle, each of three parts, namely, of calyx and corolla, both of the same bright color and delicate texture. In the Spiderwort and Trillium (Fig. 189) the three outer leaves, or sepals, are green, and different in texture from the three inner, or the petals; the stamens are six (namely, two sets of three each), and the pistils three, though partly grown together into one mass.

190

247. **Alternation of Parts.** The symmetry of the flower is likewise shown in the arrangement or relative position of successive parts. The rule is, that the parts of successive circles *alternate* with one another. That is, the petals 'stand over the intervals between the

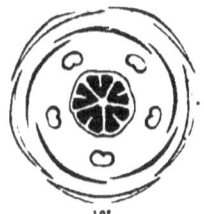

191

sepals; the stamens, when of the same number, stand over the intervals between the petals; or when twice as many, as in the Trillium, the outer set alternates with the petals, and the inner set, alternating with the other, of course stands before the petals; and the pistils alternate with these. This is shown in Fig. 189, and in the diagram, or cross-section of the same in the bud, Fig. 190. And Fig. 191 is a similar diagram or ground-plan (in the form of a

FIG. 189. Flower of Trillium erectum, or Birthroot, spread out a little, and viewed from above.

FIG. 190. Diagram or ground-plan of the same, as it would appear in a cross-section of the bud; — the parts all in the same relative position

FIG. 191. Diagram, or ground-plan, of the Flax-flower, Fig. 174.

section made across the bud) of the Flax blossom, the example of a
pattern symmetrical flower taken at the beginning of this Lesson,
with its parts all in fives.

248. Knowing in this way just the position which each organ
should occupy in the flower, it is readily understood that flowers
often become unsymmetrical through the loss of some parts, which

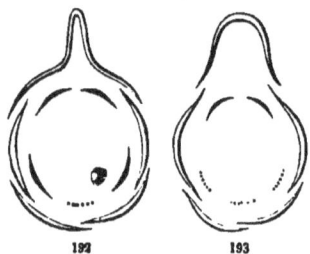

192   193

belong to the plan, but are obliterated
or left out in the execution. For example, in the Larkspur (Fig. 183,
184), as there are five sepals, there
should be five petals likewise. We
find only four; but the vacant place
where the fifth belongs is plainly recognized at the lower side of the flower.

Also the similar plan of the Monkshood (Fig. 186) equally calls for
five petals; but three of them are entirely obliterated, and the two
that remain are reduced to slender bodies, which look as unlike ordinary petals as can well be imagined. Yet their position, answering to the intervals between the upper sepals and the side ones,
reveals their true nature. All this may perhaps be more plainly
shown by corresponding diagrams of the calyx and corolla of the
Larkspur and Monkshood (Fig. 192, 193), in which the places of
the missing petals are indicated by faint dotted lines. The obliteration of stamens is a still more common case. For example, the
Snapdragon, Foxglove, Gerardia, and almost all flowers of the
large Figwort family they belong to, have the parts of the calyx
and corolla five each, but only four stamens (Fig. 194); the place
on the upper side of the flower where the fifth stamen belongs is
vacant. That there is in such cases a real obliteration of the missing part is shown by the

249. Abortive Organs, or vestiges which are sometimes met with;
— bodies which stand in the place of an organ, and represent it,
although wholly incapable of fulfilling its office. Thus, in the Figwort family, the fifth stamen, which is altogether missing in Gerardia
(Fig. 194) and most others, appears in the Figwort as a little scale,
and in Pentstemon (Fig. 195) and Turtlehead as a sort of filament
without any anther; — a thing of no use whatever to the plant, but

FIG. 192. Diagram of the calyx and corolla of a Larkspur. 193. Similar diagram of
Monkshood. The dotted lines show where the petals are wanting; one in the former, three
in the latter.

very interesting to the botanist, since it completes the symmetry of the blossom. And to show that this really is the lost stamen, it now and then bears an anther, or the rudiment of one. So the flower of Catalpa should likewise have five stamens; but we seldom find more than two good ones. Still we may generally discern the three others, as vestiges or half-obliterated stamens (Fig. 196). In separated flowers the rudiments of pistils are often found in the sterile blossom, and rudimentary stamens in the fertile blossom, as in Moonseed (Fig. 177).

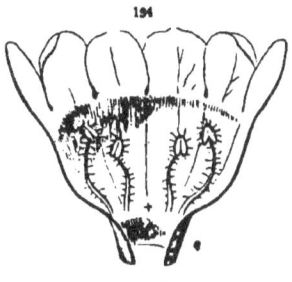

250. **Multiplication of Parts.** Quite in the opposite way, the simple plan of the flower is often more or less obscured by an increase in the number of parts. In the White Water-Lily, and in many Cactus-flowers (Fig. 197), all the parts are very numerous, so that it is hard to say upon what number the blossom is constructed. But more commonly some of the sets are few and definite in the number of their parts. The Buttercup, for instance, has five sepals and five petals, but many stamens and pistils; so it is built upon the plan of five. The flowers of Magnolia have indefinitely numerous stamens and pistils, and rather numerous floral

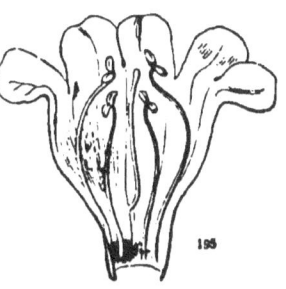

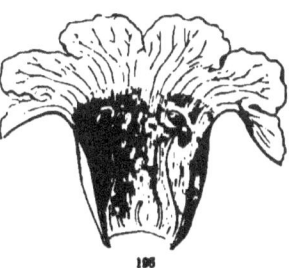

envelopes; but these latter are plainly distinguishable into sets of three; namely, there are three sepals, and six petals in two circles, or nine in three circles, — showing that these blossoms are constructed on the number three.

FIG. 194. Corolla of a purple Gerardia laid open, showing the four stamens; the cross shows where the fifth stamen would be, if present.

FIG. 195. Corolla, laid open, and stamens of Pentstemon grandiflorus of Iowa, &c., with a sterile filament in the place of the fifth stamen, and representing it.

FIG. 196. Corolla of Catalpa laid open, displaying two good stamens and three abortive vestiges of stamens.

LESSON XIV.

MORPHOLOGY OF THE FLOWER.

251. In all the plant till we came to the blossom we found nothing but root, stem, and leaves (23, 118). However various or strange their shapes, and whatever their use, everything belongs to one of these three organs, and everything above ground (excepting the rare case of aerial roots) is either stem or leaf. We discern the stem equally in the stalk of an herb, the trunk and branches of a tree, the trailing or twining Vine, the straw of Wheat or other Grasses, the columnar trunk of Palms (Fig. 47), in the flattened joints of the Prickly-Pear Cactus, and the rounded body of the Melon Cactus (Fig. 76). Also in the slender runners of the Strawberry, the tendrils of the Grape-vine and Virginia Creeper, the creeping subterranean shoots of the Mint and Couchgrass, the tubers of the Potato and Artichoke, the solid bulb of the Crocus, and the solid part or base of scaly bulbs; as is fully shown in Lesson 6. And in Lesson 7 and elsewhere we have learned to recognize the leaf alike in the thick seed-leaves of the Almond, Bean, Horsechestnut, and the like (Fig. 9 – 24), in the scales of buds (Fig. 77), and the thickened

FIG. 197. A Cactus-flower, viz. of Mamillaria cæspitosa of the Upper Missouri

scales of bulbs (Fig. 73 – 75), in the spines of the Barberry and the tendrils of the Pea, in the fleshy rosettes of the Houseleek, the strange fly-trap of Dionæa (Fig. 81), and the curious pitcher of Sarracenia (Fig. 79).

252. Now the student who understands these varied forms or *metamorphoses* of the stem and leaf, and knows how to detect the real nature of any part of the plant under any of its disguises, may readily trace the leaf into the blossom also, and perceive that, as to their morphology,

253. **Flowers are altered Branches,** and their parts, therefore, altered leaves. That is, certain buds, which might have grown and lengthened into a leafy branch, do, under other circumstances and to accomplish other purposes, develop into blossoms. In these the axis remains short, nearly as it is in the bud; the leaves therefore remain close together in sets or circles; the outer ones, those of the calyx, generally partake more or less of the character of foliage; the next set are more delicate, and form the corolla, while the rest, the stamens and pistils, appear under forms very different from those of ordinary leaves, and are concerned in the production of seed. This is the way the scientific botanist views a flower; and this view gives to Botany an interest which one who merely notices the shape and counts the parts of blossoms, without understanding their plan, has no conception of.

254. That flowers answer to branches may be shown first from their position. As explained in the Lesson on Inflorescence, flowers arise from the same places as branches, and from no other; flower-buds, like leaf-buds, appear either on the summit of a stem, that is, as a terminal bud, or in the axil of a leaf, as an axillary bud (196). And at an early stage it is often impossible to foretell whether the bud is to give rise to a blossom or to a branch.

255. That the sepals and petals are of the nature of leaves is evident from their appearance; persons who are not botanists commonly call them the leaves of the flower. The calyx is most generally green in color, and foliaceous (leaf-like) in texture. And though the corolla is rarely green, yet neither are proper leaves always green. In our wild Painted-Cup, and in some scarlet Sages, common in gardens, the leaves just under the flowers are of the brightest red or scarlet, often much brighter-colored than the corolla itself. And sometimes (as in many Cactuses, and in Carolina Allspice) there is such a regular gradation from the last leaves of the

plant (bracts or bractlets) into the leaves of the calyx, that it is impossible to say where the one ends and the other begins. And if sepals are leaves, so also are petals; for there is no clearly fixed limit between them. Not only in the Carolina Allspice and Cactus (Fig. 197), but in the Water-Lily (Fig. 198) and a variety of flowers with more than one row of petals, there is such a complete transition between calyx and corolla that no one can surely tell how many of the leaves belong to the one and how many to the other.

256. It is very true that the calyx or the corolla often takes the form of a cup or tube, instead of being in separate pieces, as in Fig. 194 – 196. It is then composed of two or more leaves grown together. This is no objection to the petals being leaves; for the same thing takes place with the ordinary leaves of many plants, as, for instance, in the upper ones of Honeysuckles (Fig. 132).

257. That stamens are of the same general nature as petals, and therefore a modification of leaves, is shown by the gradual transitions that occur between the one and the other in many blossoms; especially in cultivated flowers, such as Roses and Camellias, when they begin to *double,* that is, to change their stamens into petals. Some wild and natural flowers show the same interesting transitions. The Carolina Allspice and the White Water-Lily exhibit complete gradations not only between sepals and petals, but between petals and stamens. The sepals of the Water-Lily are green outside, but white and petal-like on the inside; the petals, in many rows, gradually grow narrower towards the centre of the flower; some of these are tipped with a trace of a yellow anther, but still are petals; the next are more contracted and stamen-like, but with a flat petal-like filament; and a further narrowing of this completes the genuine stamen. A series of these stages is shown in Fig. 198.

258. Pistils and stamens now and then change into each other in some Willows; pistils often turn into petals in cultivated flowers; and in the Double Cherry they occasionally change directly into small green leaves. Sometimes a whole blossom changes into a cluster of green leaves, as in the "green roses" which are occasionally noticed in gardens, and sometimes it degenerates into a leafy branch. So the botanist regards pistils also as answering to leaves. And his idea of a pistil is, that it consists of a leaf with its margins curved inwards till they meet and unite to form a closed cavity, the ovary, while the tip is prolonged to form the style and bear the stigma; as will be illustrated in the Lesson upon the Pistil.

259. Moreover, the arrangement of the parts of the flower answers to that of leaves, as illustrated in Lesson 10, — either to a succession of whorls alternating with each other in the manner of whorled leaves, or in some regular form of spiral arrangement.

196

LESSON XV.

MORPHOLOGY OF THE CALYX AND COROLLA.

260. HAVING studied the flower as a whole, we proceed to consider more particularly its several parts, especially as to the principal differences they present in different plants. We naturally begin with the *leaves of the blossom*, namely, the calyx and corolla. And first as to

261. **The Growing together of Parts.** It is this more than anything else which prevents one from taking the idea, at first sight, that the flower is a sort of very short branch clothed with altered leaves. For most blossoms we meet with have some of their organs grown together more or less. We have noticed it as to the corolla of Gerardia, Catalpa, &c. (Fig. 194 – 196), in Lesson 13. This growing

---

FIG. 196. Succession of sepals, petals, gradations between petals and stamens, and true stamens, of the Nymphæa, or White Water-Lily.

together takes place in two ways : either parts of the same kind,
or parts of different kinds, may be united.   The first we may call
simply the *union,* the second the *consoli-*
dation, of parts.

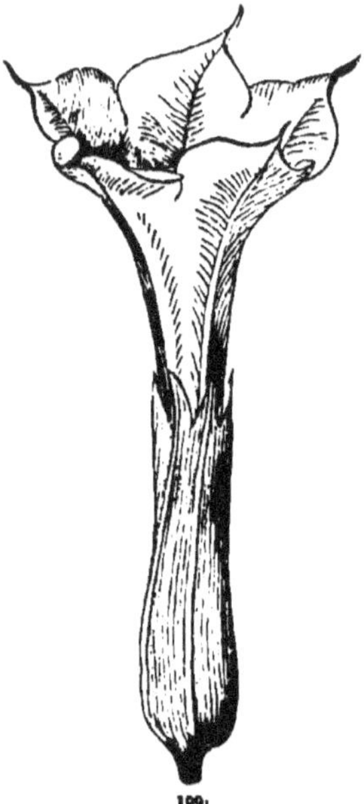

262. **Union or Cohesion** *with one another*
of parts of the same sort.   We very commonly find that the calyx or the corolla
is a cup or tube, instead of a set of leaves.
Take, for example, the flower of the Stramonium or Thorn-Apple, where both the
calyx and the corolla are so (Fig. 199);
likewise the common Morning-Glory, and
the figures 201 to 203, where the leaves
of the corolla are united into one piece,
but those of the calyx are separate.   Now
there are numerous cases of real leaves
growing together much in the same
way, — those of the common Thorough-
wort, and the upper pairs in Woodbines
or Honeysuckles, for example (Fig. 132);
so that we might expect it to occur in
the leaves of the blossom also.   And that this is the right view to
take of it plainly appears from the transitions everywhere met with
in different plants, between a calyx or a corolla of separate pieces
and one forming a perfect tube or cup.   Figures 200 to 203 show
one complete set of such gradations in the corolla, and Fig. 204 to
206 another, in short and open corollas.   How many leaves or petals
each corolla is formed of may be seen by the number of points or
tips, or of the notches (called *sinuses*) which answer to the intervals between them.

263. When the parts are united in this way, whether much or
little, the corolla is said to be *monopetalous,* and the calyx *mono-*
sepalous.   These terms mean "of one petal," or "of one sepal";
that is, of one piece.   Wherefore, taking the corolla or the calyx
as a whole, we say that it is *parted* when the parts are separate
almost to the base, as in Fig. 204 ; *cleft* or *lobed* when the notches
do not extend below the middle or thereabouts, as in Fig. 205 ;

FIG. 199.   Flower of the common Stramonium ; both the calyx and the corolla with their
parts united into a tube.

toothed or *dentate*, when only the tips are separate as short points *entire*, when the border is even, without points or notches, as in the

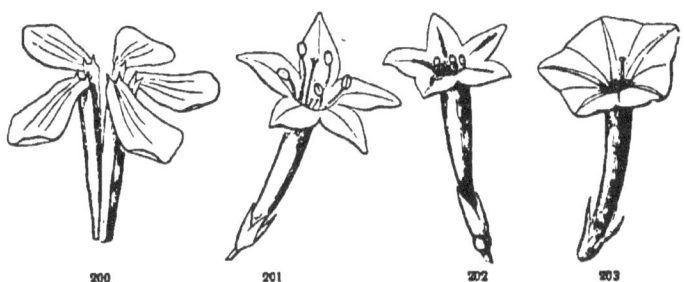

common Morning-Glory, and very nearly so in Fig. 203; and so on; — the terms being just the same as those applied to leaves and all other flat bodies, and illustrated in Lessons 8 and 9.

264. There is a set of terms applied particularly to calyxes, corollas, or other such bodies of one piece, to express their general shape, which we see is very various. The following are some of the principal : —

Wheel-shaped, or *rotate ;* when spreading out at once, without a tube or with a very short one, something in the shape of a wheel or of its diverging spokes, as in the corolla of the Potato and Bittersweet (Fig. 204, 205).

Salver-shaped, or *salver-form ;* when a flat-spreading border is raised on a narrow tube, from which it diverges at right angles,

like the salver represented in old pictures, with a slender handle beneath. The corolla of the Phlox (Fig. 208) and of the Cypress-Vine (Fig. 202) are of this sort.

FIG. 200. Corolla of Soapwort (the same in Pinks, &c.), of 5 separate, long-clawed petals.
FIG. 201. Flower of Gilia or Ipomopsis coronopifolia ; the parts answering to the claws of the petals of the last figure here all united into a tube.
FIG. 202. Flower of the Cypress-Vine ; the petals a little farther united into a five-lobed spreading border.
FIG. 203. Flower of the small Scarlet Morning-Glory, the five petals it is composed of perfectly united into a trumpet-shaped tube, with the spreading border nearly even (or entire).
FIG. 204. Wheel-shaped and five-parted corolla of Bittersweet (Solanum Dulcamara).
FIG. 205. Wheel-shaped and five-cleft corolla of the common Potato.
FIG. 206. Almost entire and very open bell-shaped corolla of a Ground Cherry (Physalis)

Bell-shaped, or *campanulate ;* where a short and broad tube widens upward, in the shape of a bell, as in Fig. 207.

Funnel-shaped, or *funnel-form ;* gradually spreading at the summit of a tube which is narrow below, in the shape of a funnel or tunnel, as in the corolla of the common Morning-Glory, and of the Stramonium (Fig. 199).

Tubular ; when prolonged into a tube, without much spreading at the border, as in the corolla of the Trumpet Honeysuckle, the calyx of Stramonium (Fig. 199), &c.

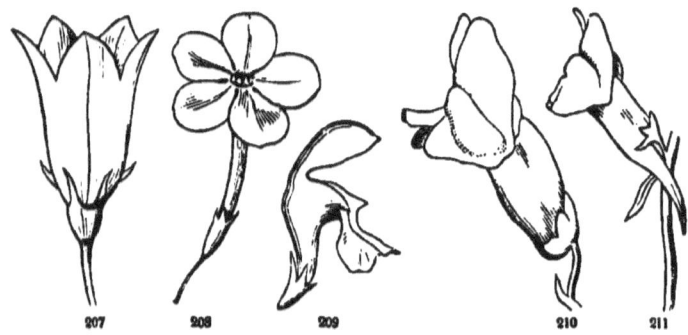

207   208   209   210   211

265. In most of these cases we may distinguish two parts; namely, the *tube*, or the portion all in one piece and with its sides upright or nearly so; and the *border* or *limb*, the spreading portion or summit. The limb may be entire, as in Fig. 203, but it is more commonly *lobed*, that is, partly divided, as in Fig. 202, or *parted* down nearly to the top of the tube, as in Fig. 208, &c.

266. So, likewise, a separate petal is sometimes distinguishable into two parts; namely, into a narrowed base or stalk-like part (as in Fig. 200, where this part is peculiarly long), called the *claw*, and a spreading and enlarged summit, or body of the petal, called the *lamina* or *blade*.

267. When parts of the same set are not united (as in the Flax, Cherry, &c., Fig. 212 – 215), we call them *distinct*. Thus the sepals or the petals are distinct when not at all united with each other. As a calyx with sepals united into one body is called *monosepalous* (263, that is, one-sepalled), or sometimes *monophyllous*, that is, one-leaved ; so, on the other hand, when the sepals are distinct, it is said to be

FIG. 207.   Flower of the Harebell, with a campanulate or bell-shaped corolla.   208. Of a Phlox, with salver-shaped corolla.   209. Of Dead-Nettle (Lamium), with labiate *ringent* (or gaping) corolla.   210. Of Snapdragon, with labiate *personate* corolla.   211. Of Toad-Flax, with a similar corolla spurred at the base.

polysepalous, that is, composed of several or many sepals. And a corolla with distinct petals is said to be *polypetalous*.

268. **Consolidation,** *the growing together of the parts of two or more different sets.* In the most natural or pattern flower (as explained in Lessons 13 and 14), the several parts rise from the receptacle or axis in succession, like leaves upon a very short stem ; the petals just above or within the sepals, the stamens just above or within these, and then the pistils next the summit or centre. 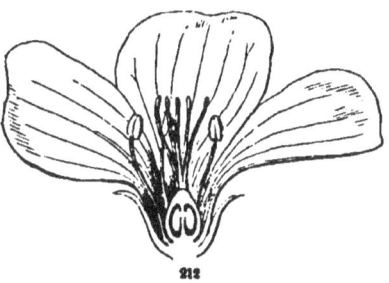 Now when contiguous parts of different sorts, one within the other, unite at their base or origin, it obscures more or less the plan of the flower, by consolidating organs which in the pattern flower are entirely separate.

269. The nature of this consolidation will be at once understood on comparing the following series of illustrations. Fig. 212 represents a flower of the common Flax, cut through lengthwise, so as to show the attachment (or what the botanist calls the *insertion*) of all the parts. Here they are all *inserted* on, that is grow out of, the receptacle or axis of the blossom. In other words, there is no union at all of the parts of contiguous circles. So the parts are said to be *free*.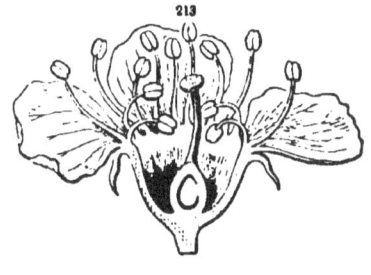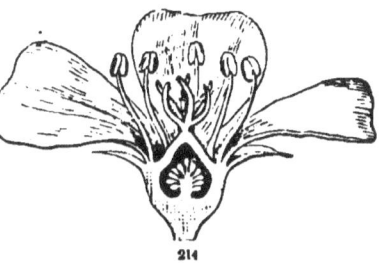

And the sepals, petals, and stamens, all springing of course from beneath the pistils, which are on the very summit of the axis, are said to be *hypogynous* (a term composed of two Greek words, meaning "under the pistil").

FIG. 212. A Flax-flower, cut through lengthwise.
FIG. 213. Flower of a Cherry, divided in the same way.
FIG. 214. Flower of the common Purslane, divided lengthwise.

270. Fig. 213 is a flower of a Cherry, cut through lengthwise in the same way. Here the petals and the stamens grow out of, that is, are *inserted* on, the calyx; in other words they cohere or are consolidated with the base of the calyx up to a certain height. In such cases they are said to be *perigynous* (from two Greek words, meaning around the pistil). The consolidation in the Cherry is confined to the calyx, corolla, and stamens: the calyx is still *free* from the pistil. One step more we have in

271. Fig. 214, which is a similar section of a flower of a Purslane.

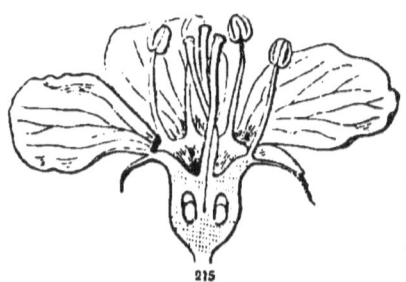

Here the lower part of the calyx (carrying with it of course the petals and stamens) is *coherent* with the surface of the whole lower half of the ovary. Therefore the calyx, seeming to rise from the middle of the ovary, is said to be *half superior*, instead of being

215

inferior, as it is when entirely free. It is better to say, however, *calyx half-adherent* to the ovary. Every gradation occurs between such a case and that of a calyx altogether *free* or inferior, as we see in different Purslanes and Saxifrages. The consolidation goes farther,

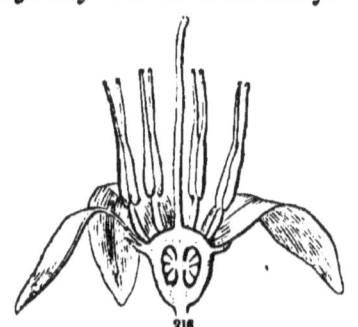

272. In the Apple, Quince, Hawthorn (Fig. 215), &c. Here the tube of the calyx is consolidated with the whole surface of the ovary; and its

216

limb, or free part, therefore appears to spring from its top, instead of underneath it, as it naturally should. So the calyx is said to be *superior*, or (more properly) *adherent* to, or *coherent* with, the ovary. In most cases (and very strikingly in the Evening Primrose), the tube of the calyx is continued on more or less beyond the ovary, and has the petals and stamens consolidated with it for some distance; these last, therefore, being borne on the calyx, are said to be *perigynous*, as before (270).

FIG. 215. Flower of a Hawthorn, divided lengthwise.
FIG. 216. Flower of the Cranberry, divided lengthwise.

273. But if the tube of the calyx ends immediately at the summit of the ovary, and its lobes as well as the corolla and stamens are as it were inserted directly on the ovary, they are said to be *epigynous* (meaning on the pistil), as in Cornel, the Huckleberry, and the Cranberry (Fig. 216).

274. **Irregularity of Parts** in the calyx and corolla has already been noticed (244) as sometimes obstructing one's view of the real plan of a flower. There is infinite variety in this respect; but what has already been said will enable the student to understand these irregularities when they occur. We have only room to mention one or two cases which have given rise to particular names. A very common kind, among polypetalous (267) flowers, is

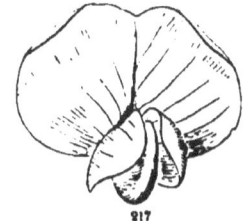

275. The *Papilionaceous* flower of the Pea, Bean, and nearly all that family. In this we have an irregular corolla of a peculiar shape, which Linnæus likened to a butterfly (whence the term, *papilio* being the Latin name for a butterfly) ; but the resemblance is not very obvious. The five petals of a papilionaceous corolla (Fig. 217) have received different names taken from widely different objects. The upper and larger petal (Fig. 218, *s*), which is generally wrapped round all the rest in the bud, is called the *standard* or *banner.* The two side petals (*w*) are called the *wings.* And the two anterior ones (*k*), the blades of which commonly stick together a little, and which en-

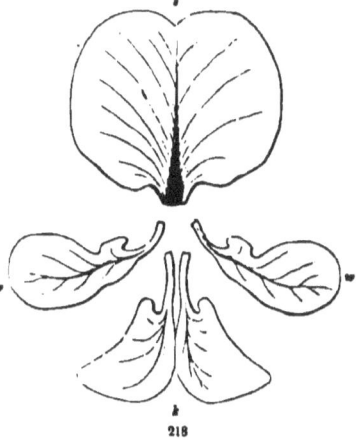

close the stamens and pistil in the flower, from their forming a body shaped somewhat like the keel, or rather the prow, of an ancient boat, are together named the *keel.*

276. The *Labiate* or *bilabiate* (that is, *two-lipped*) flower is a very common form of the monopetalous corolla, as in the Snapdragon

FIG. 217. Front view of the papilionaceous corolla of the Locust-tree. 218. The parts of the same, displayed

S & F—6

(Fig. 210), Toad-Flax (Fig. 211), Dead-Nettle (Fig. 209), Catnip,
Horsemint, &c.; and in the Sage, the Catalpa, &c., the calyx also is
two-lipped. This is owing to unequal union of the different parts of
the same sort, as well as to diversity of shape. In the corolla two
of the petals grow together higher than the rest, sometimes to the
very top, and form the *upper lip*, and the three remaining ones join
on the other side of the flower to form the *lower lip*, which therefore
is more or less three-lobed, while the upper lip is at most only two-
lobed. And if the calyx is also two-lipped, as in the Sage, — since
the parts of the calyx always alternate with those of the corolla
(247), — then the upper lip has three lobes or teeth, namely, is composed of three sepals united, while the lower has only two ; which is
the reverse of the arrangement in the corolla. So that all these
flowers are really constructed on the plan of five, and not on that of
two, as one would at first be apt to suppose. In Gerardia, &c. (Fig.
194, 195), the number five is evident in the calyx and corolla, but is
more or less obscured in the stamens (249). In Catalpa this number is masked in the calyx by irregular union, and in the stamens by
abortion. A different kind of irregular flower is seen in

277. The *Ligulate* or *strap-
shaped* corolla of most compound flowers. What was
called the compound flower
of a Dandelion, Succory (Fig.
221), Thistle, Sunflower, Aster, Whiteweed, &c., consists
of many distinct blossoms,
closely crowded together into

a head, and surrounded by an involucre (208). People who are not
botanists commonly take the whole for one flower, the involucre for
a calyx, and corollas of the outer or of all the flowers as petals.
And this is a very natural mistake when the flowers around the
edge have flat and open or strap-shaped corollas, while the rest
are regular and tubular, but small, as in the Whiteweed, Sunflower,
&c. Fig. 219 represents such a case in a Coreopsis, with the
head, or so-called compound flower, cut through ; and in Fig. 220
we see one of the perfect flowers of the centre or *disk*, with a regular tubular corolla (*a*), and with the slender bract (*b*) from whose

FIG. 219. Head of flowers (the so-called " compound flower ") of Coreopsis, divided
lengthwise.

axil it grew; and also one belonging to the margin, or *ray*, with
a strap-shaped corolla (*c*), borne in the axil of a leaf or bract of

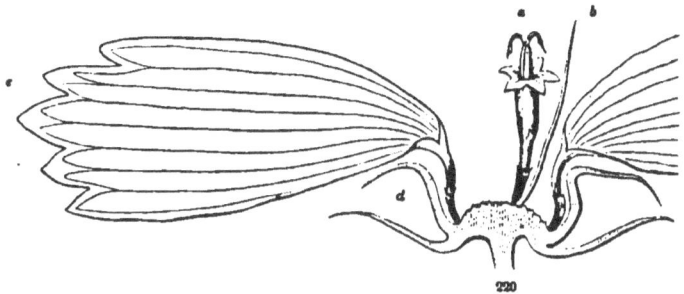

the involucre (*d*).   Here the *ray-flower* consists merely of a strap-
shaped corolla, raised on the small rudiment of an ovary; it is
therefore a *neutral* flower, like those of the ray or margin of the
cluster in Hydrangea (229, Fig. 167), only of a different shape.
More commonly the flowers with a strap-shaped corolla are *pistillate*, that is, have a pistil only, and produce seed like the others,
as in Whiteweed.   But in the Dandelion, Succory (Fig. 221, 222),

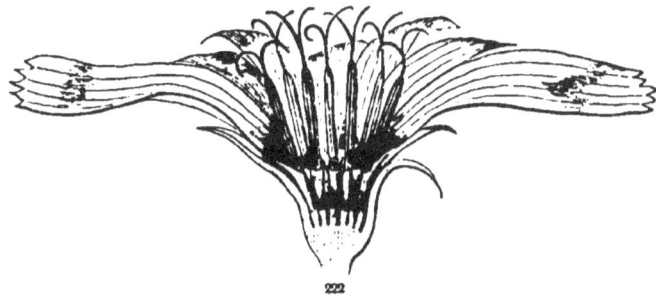

and all of that tribe, these flowers are perfect, that is, bear both
stamens and pistils.   And moreover all the flowers of the head are
strap-shaped and alike.

278.   Puzzling as these strap-shaped corollas appear at first view,
an attentive inspection will generally reveal the plan upon which
they are constructed.   We can make out pretty plainly, that each
one consists of five petals (the tips of which commonly appear as five
teeth at the extremity), united by their contiguous edges, except on

FIG. 220.   A slice of Fig. 219, more enlarged, with one tubular perfect flower (*a*) left
standing on the receptacle, with its bractlet or chaff (*b*), one ligulate, neutral ray-flower (*c*),
and part of another: *d*, section of bracts or leaves of the involucre.

FIG. 222.   Head of flowers of Succory, cut through lengthwise and enlarged.

one side, and spread out flat. To prove that this is the case, we have
only to compare such a corolla (that of Coreopsis, Fig. 220, c, or
one from the Succory, for instance) with that of the Cardinal-flower,
or of any other Lobelia, which is equally split down along one side ;
and this again with the less irregular corolla of the Woodbine, partially split down on one side.

LESSON XVI.

ÆSTIVATION, OR THE ARRANGEMENT OF THE CALYX AND CO-
ROLLA IN THE BUD.

279. ÆSTIVATION or *Præfloration* relates to the way in which
the leaves of the flower, or the lobes of the calyx or corolla, are
placed with respect to each other in the bud. This is of some
importance in distinguishing different families or tribes of plants,
being generally very uniform in each. The æstivation is best seen

FIG. 221. Compound flowers, i. e. heads of flowers, of Succory.

by making a horizontal slice of the flower-bud when just ready to open ; and it may be expressed in diagrams, as in Fig. 223, 224.

280. The pieces of the calyx or the corolla either overlap each other in the bud, or they do not. When they do not, the æstivation is commonly

Valvate, as it is called when the pieces meet each other by their abrupt edges without any infolding or overlapping ; as the calyx of the Linden or Basswood (Fig. 223) and the Mallow, and the corolla of the Grape, Virginia Creeper, &c. Or it may be

Induplicate, which is valvate with the margins of each piece projecting inwards, or involute (like the leaf in Fig. 152), as in the calyx of Virgin's-Bower and the corolla of the Potato, or else

Reduplicate, like the last, but the margins projecting outwards instead of inwards ; these last being mere variations of the valvate form.

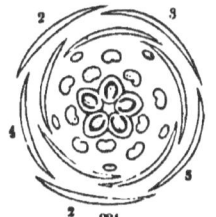

223

281. When the pieces overlap in the bud, it is in one of two ways : either every piece has one edge in and one edge out ; or some pieces are wholly outside and others wholly inside. In the first case the æstivation is

Convolute or *twisted*, as in the corolla of Geranium (most commonly, Fig. 224), Flax (Fig. 191), and of the Mallow Family: Here one edge of every petal covers the next before it, while its other edge is covered by the next behind it. In the second case it is

Imbricated or *imbricate*, or *breaking joints*, like shingles on a roof, as in the calyx of Geranium (Fig. 224) and of Flax (Fig. 191), and the corolla of the Linden (Fig. 223). In these cases the parts are five in number ; and the regular way then is (as in the calyx of the figures above cited) to have two pieces entirely external (1 and 2), one (3) with one edge covered by the first, while the other edge covers that of the adjacent one on the other side, and two (4 and 5) wholly within, their margins at least being covered by the rest. That is, they just represent a circle of five leaves spirally arranged on the five-ranked or $\frac{2}{5}$ plan (187, 188, and Fig. 143 – 145), only with the stem shortened so as to bring the parts close together. The spiral arrangement of the parts of

FIG. 223. Section across the flower-bud of Linden.
FIG. 224. Section across the flower-bud of Geranium : the sepals numbered in their order

the blossom is the same as that of the foliage, — an additional evidence that the flower is a sort of branch. The petals of the Linden, with only one outside and one inside, as shown in Fig. 223, exhibit a gradation between the imbricated and the convolute modes. When the parts are four in number, generally two opposite ones overlap the other two by both edges. When three in number, then one is outermost, the next has one edge out and the other covered, and the third is within, being covered by the other two; as in Fig. 190. This is just the three-ranked ($\frac{1}{3}$) spiral arrangement of leaves (186, and Fig. 171).

282. In the Mignonette, and some other flowers, the æstivation is *open;* that is, the calyx and corolla are not closed at all over the other parts of the flower, even in the young bud.

283. When the calyx or the corolla is tubular, the shape of the tube in the bud has sometimes to be considered, as well as the way the lobes are arranged. For example, it may be

Plaited or *plicate,* that is, folded lengthwise; and the plaits may either be turned outwards, forming projecting ridges, as in the corolla of Campanula ; or turned inwards, as in the corolla of the Gentian, &c. When the plaits are wrapped round all in one direction, so as to cover one another in a convolute manner, the æstivation is said to be

Supervolute, as in the corolla of Stramonium (Fig. 225) and the Morning-Glory; and in the Morning-Glory it is twisted besides.

FIG. 225. Upper part of the corolla of a Stramonium (Datura meteloides), in the bud. Underneath is a cross-section of the same.

LESSON XVII.

MORPHOLOGY OF THE STAMENS.

284. THE STAMENS exhibit nearly the same kinds of variation in different species that the calyx and corolla do. They may be *distinct* (that is, separate from each other, 267) or united. They may be *free* (269), or else *coherent* with other parts : this concerns

285. **Their Insertion,** or place of attachment, which is most commonly the same as that of the corolla. So, stamens are

Hypogynous (269), when they are borne on the receptacle, or axis of the flower, under the pistils, as they naturally should be, and as is shown in Fig. 212.

Perigynous, when borne on (that is coherent below with) the calyx ; as in the Cherry, Fig. 213.

Epigynous, when borne on the ovary, apparently, as in Fig. 216. To these we may add

Gynandrous (from two Greek words, answering to "stamens and pistil united"), when the stamens are consolidated with the style, so as to be borne by it, as in the Lady's Slipper (Fig. 226) and all the Orchis Family. Also

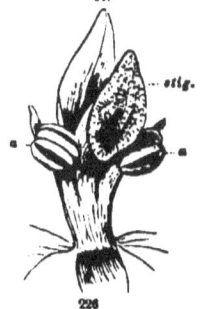

Epipetalous (meaning on the petals), when they are borne by the corolla ; as in Fig. 194, and in most monopetalous blossoms. As to

286. **Their Union with each other,** the stamens may be united by their filaments or by their anthers. In the former case they are

Monadelphous (from two Greek words, meaning " in one brotherhood "), when united by their filaments into one set, usually into a ring or cup below, or into a tube, as in the Mallow Family, the Passion-flower, and the Lupine. (Fig. 228).

Diadelphous (in two brotherhoods), when so united in two sets, as in the Pea and almost all papilionaceous flowers (275) : here the stamens are nine in one set, and one in the other (Fig. 227).

FIG. 226. Style of a Lady's Slipper (Cypripedium), and stamens united with it : *a, a,* the anthers of the two good stamens ; *st.,* an abortive stamen, what should be its anther changed into a petal-like body ; *stig.,* the stigma.

Triadelphous, in three sets or parcels, as in the common St. Johns-wort ; or

Polyadelphous, when in more numerous sets, as in the Loblolly

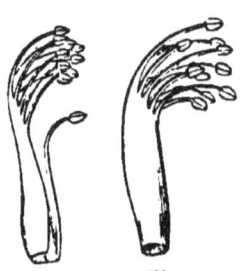

Bay, where they are in five clusters. On the other hand, stamens are said to be

Syngenesious, when united by their an-thers (Fig. 229, 230), as they are in Lobelia, in the Violet (slightly), and in what are called *compound flowers*, such as the Thistle, Sunflower, Coreopsis (Fig. 220), and Suc-cory (Fig. 222). In Lobelia, and in the Squash and Pumpkin, the stamens are united both by their anthers and their filaments.

287. **Their Number** in the flower is sometimes expressed by terms compounded of the Greek numerals and the word used to signify stamen ; as, *monandrous*, for a flower having only one stamen ; *diandrous*, one with two stamens ; *triandrous*, with three stamens ; *te-trandrous*, with four stamens ; *pentandrous*, with five stamens ; and so on, up to *polyan-drous* (meaning with many stamens), when there are twenty or a larger number, as in a Cactus (Fig. 197). All such terms may be found in the Glossary at the end of the book.

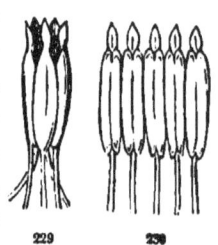

288. Two terms are used to express particular numbers with un-equal length. Namely, the stamens are *didynamous* when only four in number, two longer than the other two, as in the Mint, Catnip, Gerardia (Fig. 194), Trumpet-Creeper, &c. ; and *tetradynamous*, when they are six, with four of them regularly longer than the other two, as in Mustard (Fig. 188), and all that family.

289. **Their Parts.** As already shown (233), a stamen consists of two parts, the *Filament* and the *Anther* (Fig. 231).

290. **The Filament** is a kind of stalk to the anther : it is to the anther nearly what the petiole is to the blade of a leaf. Therefore it is not an essential part. As a leaf may be without a stalk, so the anther may be *sessile*, or without a filament. When present,

FIG. 227. Diadelphous stamens of the Pea, &c. 228. Monadelphous stamens of the Lupine.

FIG. 229. Syngenesious stamens of Coreopsis (Fig. 220, *e*), &c. 230. Same, with the tube of anthers split down on one side and spread open.

the filament may be of any shape ; but it is commonly thread-like, as in Fig. 231, 234, &c.

291. **The Anther** is the essential part of the stamen. It is a sort of case, filled with a fine powder, called *Pollen*, which serves to fertilize the pistil, so that it may perfect seeds. The anther may be considered, first, as to

292. **Its Attachment** to the filament. Of this there are three ways ; namely, the anther is

Innate (as in Fig. 232), when it is attached by its base to the very apex of the filament, turning neither inwards nor outwards ; or

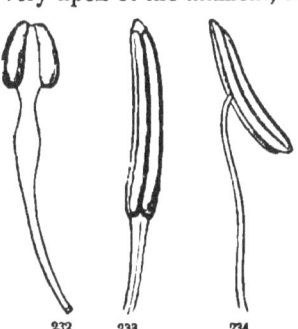

Adnate (as in Fig. 233), when attached by one face, usually for its whole length, to the side of the filament ; and

Versatile (as in Fig. 234), when fixed by its middle only to the very point of the filament, so as to swing loosely, as we see it in the Lily, in Grasses, &c.

293. In both the last-named cases, the anther either looks inwards or outwards. When it is turned inwards, or is fixed to that side of the filament which looks towards the pistil or centre of the flower, the anther is *incumbent* or *introrse*, as in Magnolia and the Water-Lily. When turned outwards, or fixed to the outer side of the filament, it is *extrorse*, as in the Tulip-tree.

294. **Its Structure,** &c. There are few cases in which the stamen bears any resemblance to a leaf. Nevertheless, the botanist's idea of a stamen is, that it answers to a leaf developed in a peculiar form and for a special purpose. In the filament he sees the stalk of the leaf ; in the anther, the blade. The blade of a leaf consists of two similar sides ; so the anther consists of two lobes or cells, one answering to the left, the other to the right, side of the blade. The two lobes are often connected by a prolongation of the filament, which answers to the midrib of a leaf · this is called the *connective*. It is very conspicuous in Fig. 232, where the connective is so broad that it separates the two cells of the anther to some distance from each other.

FIG. 231. A stamen : *a*, filament ; *b*, anther discharging pollen.
FIG. 232. Stamen of Isopyrum, with innate anther. 233. Of Tulip-tree, with adnate (and extrorse) anther. 234 Of Evening Primrose, with versatile anther.

10 *

295. To discharge the pollen, the anther opens (or is *dehiscent*)

at maturity, commonly by a line along the whole length of each cell, and which answers to the margin of the leaf (as in Fig. 231); but when the anthers are extrorse, this line is often on the outer face, and when introrse, on the inner face of each cell. Sometimes the anther opens only by a chink, hole, or pore at the top, as in the Azalea, Pyrola or False Wintergreen (Fig. 235), &c.; and sometimes a part of the face separates as a sort of trap-door (or valve), hinged at the top, and opening to allow the escape of the pollen, as in the Sassafras, Spice-bush, and Barberry (Fig. 236). Most anthers are really four-celled when young; a slender partition running lengthwise through each cell and dividing it into two compartments, one answering to the upper, and the other to the lower, layer of the green pulp of the leaf. Occasionally the anther becomes one-celled. This takes place mostly by *confluence*, that is, the two cells running together into one, as they do

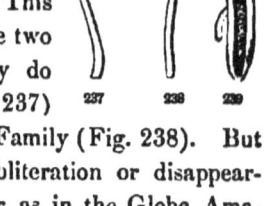

slightly in Pentstemon (Fig. 237) and thoroughly in the Mallow Family (Fig. 238). But sometimes it occurs by the obliteration or disappearance of one half of the anther, as in the Globe Amaranth of the gardens (Fig. 239).

296. The way in which a stamen is supposed to be constructed out of a leaf, or rather on the plan of a

leaf, is shown in Fig. 240, an ideal figure, the lower part representing a stamen with the top of its anther cut away; the upper, the corresponding upper part of a leaf. — The use of the anther is to produce

297. Pollen. This is the powder, or fine dust, commonly of a yellow color, which fills the cells of the anther, and is discharged during blossoming, after which the stamens generally fall off or wither away.

FIG. 235. Stamen of Pyrola; the anther opening by holes at the top.
FIG. 236. Stamen of Barberry; the anther opening by uplifted valves.
FIG. 237. Stamen of Pentstemon pubescens; anther-cells slightly confluent.
FIG. 238. Stamen of Mallow; the two cells confluent into one, opening round the margin.
FIG. 239. Anther of Globe Amaranth, of only one cell; the other cell wanting.
FIG. 240. Diagram of the lower part of an anther, cut across above, and the upper part of a leaf, to show how the one answers to the other.

Under the microscope it is found to consist of grains, usually round or oval, and all alike in the same species, but very different in different plants. So that the plant may sometimes be recognized from the pollen alone.

298. A grain of pollen is made up of two coats ; the outer coat thickish, but weak, and frequently adorned with lines or bands, or studded with points ; the inner coat is extremely thin and delicate, but extensible, and its cavity is filled with a thickish fluid, often rendered turbid by an immense number of minute grains that float in it. When wet, the grains absorb the water and swell so much that many kinds soon burst and discharge their contents.

299. Figures 241 – 250 represent some common sorts of pollen, magnified one or two hundred diameters, viz. : — A pollen-grain of the Musk Plant, spirally grooved. One of Sicyos, or One-seeded Cucumber, beset with bristly points and marked by smooth bands. One of the Wild Balsam-Apple (Echinocystis), grooved lengthwise. One of Hibiscus or Rose-Mallow, studded with prickly points. One of Succory, many-sided, and dotted with fine points. A grain of the curious compound pollen of Pine. One from the Lily, smooth and oval. One from Enchanter's Nightshade, with three small lobes on the angles. Pollen of Kalmia, composed of four grains united, as in all the Heath family. A grain from an Evening Primrose, with a central body and three large lobes. The figures number from left to right, beginning at the top.

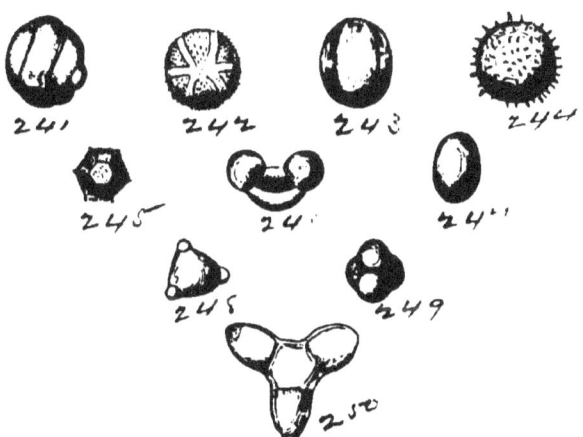

LESSON XVIII.

MORPHOLOGY OF PISTILS.

300. THE PISTIL, when only one, occupies the centre of the flower; when there are two pistils, they stand facing each other in the centre of the flower; when several, they commonly form a ring or circle; and when very numerous, they are generally crowded in rows or spiral lines on the surface of a more or less enlarged or elongated receptacle.

301. Their number in a blossom is sometimes expressed, in Systematic Botany, by terms compounded of the Greek numerals and the Greek word used to signify pistil, in the following way. A flower with one pistil is said to be *monogynous ;* with two, *digynous ;* with three, *trigynous ;* with four, *tetragynous ;* with five, *pentagynous,* and so on ; with many pistils, *polygynous,* — terms which are explained in the Glossary, but which there is no need to commit to memory.

302. **The Parts of a Pistil,** as already explained (234), are the *Ovary,* the *Style,* and the *Stigma.* The ovary is one essential part: it contains the rudiments of seeds, called *Ovules.* The stigma at the summit is also essential: it receives the pollen, which fertilizes the ovules in order that they may become seeds. But the style, the tapering or slender column commonly borne on the summit of the ovary, and bearing the stigma on its apex or its side, is no more necessary to a pistil than the filament is to the stamen. Accordingly, there is no style in many pistils: in these the stigma is *sessile,* that is, rests directly on the ovary. The stigma is very various in shape and appearance, being sometimes a little knob (as in the Cherry, Fig. 213), sometimes a small point, or small surface of bare, moist tissue (as in Fig. 254–256), and sometimes a longitudinal crest or line (as in Fig. 252, 258, 267, 269), and also exhibiting many other shapes.

303. The pistil exhibits an almost infinite variety of forms, and many complications. To understand these, it is needful to begin with the simple kinds, and to proceed gradually to the complex. And, first of all, the student should get a clear notion of

304. **The Plan or Ideal Structure of the Pistil,** or, in other words, of the way in which a simple pistil answers to a leaf. Pistils are either

simple or *compound.* A simple pistil answers to a single leaf. A compound pistil answers to two or more leaves combined, just as a monopetalous corolla (263) answers to two or more petals, or leaves of the flower, united into one body. In theory, accordingly,

305. **The Simple Pistil, or Carpel** (as it is sometimes called), consists of the blade of a leaf, curved until the margins meet and unite, forming in this way a closed case or pod, which is the ovary. So that the upper face of the altered leaf answers to the inner surface of the ovary, and the lower, to its outer surface. And the ovules are borne on what answers to the united edges of the leaf. The tapering summit, rolled together and prolonged, forms the style, when there is any; and the edges of the altered leaf turned outwards, either at the tip or along the inner side of the style, form the stigma. To make this perfectly clear, compare a leaf folded together in this way (as in Fig. 251) with a pistil of a Garden Pæony, or Larkspur, or with that in Fig. 252; or, later in the season, notice how these, as ripe pods, split down along the line formed by the united edges, and open out again into a sort of leaf, as in the Marsh-Marigold (Fig. 253). In the Double-flowering Cherry the pistil occasion ally is found changed back again into a small green leaf, partly folded, much as in Fig. 251.

251   252   253

306. Fig. 172 represents a simple pistil on a larger scale, the ovary cut through to show how the ovules (when numerous) are attached to what answers to the two margins of the leaf. The Stonecrop (Fig. 168) has five such pistils in a circle, each with the side where the ovules are attached turned to the centre of the flower.

307. The line or seam down the inner side, which answers to the united edges of the leaf, and bears the ovules, is called the *ventral* or *inner Suture.* A corresponding line down the back of the ovary, and which answers to the middle of the leaf, is named the *dorsal* or *outer Suture.*

308. The ventral suture inside, where it projects a little into the

FIG. 251.   A leaf rolled up inwards, to show how the pistil is supposed to be formed.
FIG. 252.   Pistil of Isopyrum biternatum cut across, with the inner suture turned towards the eye.
FIG. 253.   Pod or ripe pistil of the Caltha, or Marsh-Marigold, after opening.

cavity of the ovary, and bears the ovules, is called the *Placenta*. Obviously a simple pistil can have but one placenta; but this is in its nature double, one half answering to each margin of the leaf. And if the ovules or seeds are at all numerous, they will be found to occupy two rows, one for each margin, as we see in Fig. 252, 172, in the Marsh-Marigold, in a Pea-pod, and the like.

309. A simple pistil obviously can have but one cavity or cell; except from some condition out of the natural order of things. But the converse does not hold true: all pistils of a single cell are not simple. Many compound pistils are one-celled.

310. A simple pistil necessarily has but one style. Its stigma, however, may be double, like the placenta, and for the same reason (305); and it often exhibits two lines or crests, as in Fig. 252, or it may even be split into two lobes.

311. **The Compound Pistil** consists of two, three, or any greater

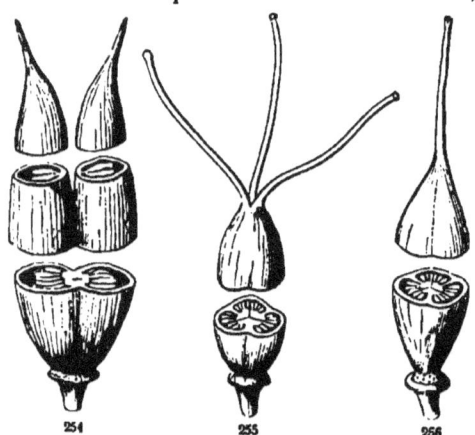

number of pistil-leaves, or carpels (305), in a circle, united into one body, at least by their ovaries. The Cultivated Flax, for example (Fig. 212), has a compound pistil composed of five simple ones with their ovaries united, while the five styles are separate. But in one of our

254   255   256

wild species of Flax, the styles are united into one also, for about half their length. So the Common St. John's-wort of the fields has a compound ovary, of three united carpels, but the three styles are separate (Fig. 255), while some of our wild, shrubby species have the styles also combined into one (Fig. 256), although in the fruit they often split into three again. Even the ovaries may only partially combine with each other, as we see in different species of Saxifrage, some having their two pistils nearly separate, while in others they

FIG. 254. Pistil of a Saxifrage, of two simple carpels or pistil-leaves, united at the base only, cut across both above and below.

FIG. 255. Compound pistil of common St. John's-wort, cut across: styles separate.

FIG. 256. The same of shrubby St. John's-wort; the three styles united into one.

are joined at the base only, or else below the middle (as in Fig. 254), and in some they are united quite to the top.

312. Even when the styles are all consolidated into one, the stigmas are often separate, or enough so to show by the number of their lobes how many simple pistils are combined to make the compound one. In the common Lily, for instance, the three lobes of the stigma, as well as the three grooves down the ovary, plainly tell us that the pistil is made of three combined. But in the Day-Lily the three lobes of the stigma are barely discernible by the naked eye, and in the Spiderwort (Fig. 257) they are as perfectly united into one as the ovaries and styles are. Here the number of cells in the ovary alone shows that the pistil is compound. These are all cases of

313. **Compound Pistils with two or more Cells,** namely, with as many cells as there are simple pistils, or carpels, that have united to compose the organ. They are just what would be formed if the simple pistils (two, three, or five in a circle, as the case may be), like those of a Pæony or Stonecrop, all pressed together in the centre of the flower, were to cohere by their contiguous parts.

314. As each simple ovary has its placenta, or seed-bearing line (308), at the inner angle, so the resulting compound ovary has as many *axile placentæ* (that is, as many placentæ in the axis or centre) as there are pistil-leaves in its composition, but all more or less consolidated into one. This is shown in the cross-sections, Fig. 254 – 256, &c.

315. The partitions (or *Dissepiments*, as they are technically named) of a compound ovary are accordingly part of the walls or the sides of the carpels which compose it. Of course they are double, one layer belonging to each carpel; and in ripe pods they often split into the two layers.

316. We have described only one, though the commonest, kind of compound pistil. There are besides

317. **One-celled Compound Pistils.** These are of two sorts, those with *axile*, and those with *parietal placentæ*. That is, first, where the ovules or seeds are borne in the axis or centre of the ovary, and, secondly, where they are borne on its walls. The first of these cases, or that

---

FIG. 257. Pistil of Spiderwort (Tradescantia): the three-celled ovary cut across.

318. **With a Free Central Placenta.** is what we find in Purslane (Fig. 214), and in most Chickweeds (Fig. 258, 259) and Pinks. The difference between this and the foregoing case is only that the delicate partitions have very early vanished; and traces of them may often be detected. Or sometimes this is a variation of the mode

319. **With Parietal Placentæ,** namely, with the ovules and seeds borne on the sides or wall (*parietes*) of the ovary. The pistil of the Prickly Poppy, Bloodroot, Violet, Frost-weed (Fig. 261), Gooseberry, and of many Hypericums, are of this sort. To understand it perfectly, we have only to imagine two, three, or any number of carpel-leaves (like that of Fig. 251), arranged in a circle, to unite by their contiguous edges, and so form one ovary or pod (as we have endeavored to show in Fig. 260); —very much as in the Stramonium (Fig. 199) the five petals unite by their edges to compose a mono-petalous corolla, and the five sepals to form a tubular calyx. Here each carpel is an open leaf, or partly open, bearing ovules along its margins; and each placenta consists of the contiguous margins of two pistil-leaves grown together.

320. All degrees occur between this and the sev-eral-celled ovary with the placentæ in the axis. Com-pare, for illustration, the common St. John's-worts, Fig. 255 and 256, with Fig. 262, a cross-section of the ovary of a different species, in which the three large placentæ meet in the axis, but scarcely unite, and with Fig. 263, a similar section of the ripe pod of the same plant, showing three parietal placentæ borne on imperfect partitions projecting a little way into the general cell. Fig. 261 is the same in plan, but with hardly any trace of partitions; that is, the united edges of the leaves only slightly project into the cell.

FIG. 258. Pistil of a Sandwort, with the ovary divided lengthwise; and 259, the same divided transversely, to show the free central placenta

FIG. 260. Plan of a one-celled ovary of three carpel-leaves, with parietal placentæ, cut across below, where it is complete; the upper part showing the top of the three leaves it is composed of, approaching, but not united.

FIG. 261 Cross-section of the ovary of Frost-weed (Helianthemum), with three parietal placentæ, bearing ovules.

321. The ovary, especially when compound, is often covered by and united with the tube of the calyx, as has already been explained (272). We describe this by saying either "ovary adherent," or "calyx adherent," &c. Or we say "*ovary inferior*," when the tube of the calyx is adherent throughout to the surface of the ovary, so that its lobes, and all the rest of the flower, appear to be borne on its summit, as in Fig. 215 and Fig. 216; or "*half-inferior*," as in the Purslane (Fig. 214), 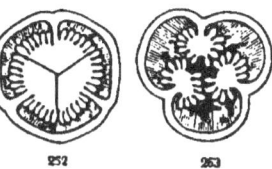 where the calyx is adherent part way up; or "*superior*," where the calyx and the ovary are not combined, as in the Cherry (Fig. 213) and the like, that is, where these parts are *free*. The term "ovary superior," therefore, means just the same as "calyx inferior"; and "ovary inferior," the same as "calyx superior."

322. **Open or Gymnospermous Pistil.** This is what we have in the whole Pine family, the most peculiar, and yet the simplest, of all pistils. While the ordinary simple pistil in the eye of the botanist represents a leaf rolled together into a closed pod (305), those of the Pine, Larch (Fig. 264), Cedar, and Arbor-Vitæ (Fig. 265, 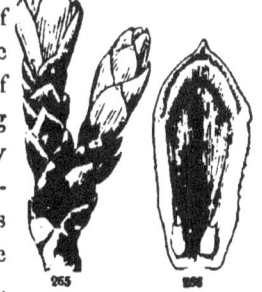 266) are plainly open leaves, in the form of scales, each bearing two or more ovules on the inner face, next the base. At the time of blossoming, these pistil-leaves of the young cone diverge, and the pollen, so abundantly shed from the staminate blossoms, falls directly upon the exposed ovules. Afterwards the scales close over each other until the seeds are ripe. Then they separate again, that the seeds may be shed. As their ovules and seeds are not enclosed in a pod, all such plants are said to be *Gymnospermous*, that is, *naked-seeded*.

FIG. 262. Cross-section of the ovary of Hypericum graveolens. 263. Similar section of the ripe pod of the same.

FIG. 264. A pistil, that is, a scale of the cone, of a Larch, at the time of flowering; inside view, showing its pair of naked ovules.

FIG. 265. Branchlet of the American Arbor-Vitæ, considerably larger than in nature, terminated by its pistillate flowers, each consisting of a single scale (an open pistil), together forming a small cone.

FIG. 266. One of the scales or pistils of the last, removed and more enlarged, the inside exposed to view, showing a pair of ovules on its base.

323. **Ovules** (234). These are the bodies which are to become seeds. They are either *sessile*, that is, stalkless, or else borne on a stalk, called the *Funiculus*. They may be produced along the whole length of the cell, or only at some part of it, generally either at the top or the bottom. In the former case they are apt to be numerous; in the latter, they may be few or single (*solitary*, Fig. 267 – 269). As to their direction, ovules are said to be

Horizontal, when they are neither turned upwards nor down-,wards, as in Fig. 252, 261;

Ascending, when rising obliquely upwards, usually from the side of the cell, not from its very base, as in the Buttercup (Fig. 267),

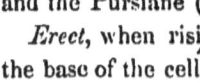

and the Purslane (Fig. 214);

Erect, when rising upright from the base of the cell, as in the Buck-wheat (Fig. 268);

Pendulous, when hanging from towards the top, as in the Flax (Fig. 212); and

Suspended, when hanging perpendicularly from the very summit of the cell, as in the Anemone (Fig. 269), Dogwood, &c. All these terms equally apply to seeds.

324. An ovule consists of a pulpy mass of tissue, the *Nucleus* or kernel, and usually of one or two coats. In the nucleus the embryo is formed, and the coats become the skin or coverings of the seed. There is a hole (*Orifice* or *Foramen*) through the coats, at the place which answers to the apex of the ovule. The part by which the ovule is attached is its base; the point of attachment, where the ripe seed breaks away and leaves a scar, is named the *Hilum*. The place where the coats blend, and cohere with each other and with the nucleus, is named the *Chalaza*. We will point out these parts in illustrating the four principal kinds of ovule. These are not difficult to understand, although ovules are usually so small that a good magnifying-glass is needed for their examination. Moreover, their names, all taken from the Greek, are unfortunately rather formidable.

325. The simplest sort, although the least common, is what is called the

Orthotropous, or *straight* ovule. The Buckwheat affords a good

FIG. 267. Section of the ovary of a Buttercup, lengthwise, showing its ascending ovule.
FIG. 2 8. Section of the ovary of Buckwheat, showing the erect ovule.
FIG 2 9. Section of the ovary of Anemone, showing its suspended ovule

instance of it : it is shown in its place in the ovary in Fig. 268, also detached in Fig. 270, and a much more magnified diagram of it in Fig. 274. In this kind, the orifice (f) is at the top, the chalaza and the hilum (c) are blended at the base or point of attachment, which is at the opposite end ; and the axis of the ovule is straight.

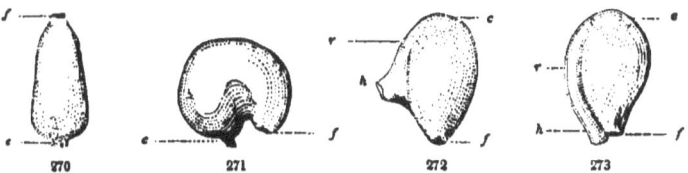

If such an ovule were to grow on one side more than on the other, and double up, or have its top pushed round as it enlarges, it would become a

Campylotropous or *curved* ovule, as in Cress and Chickweed (Fig. 271). Here the base remains as in the straight kind, but its apex with the orifice is brought round close to it. — Much the most common form of all is the

Anatropous or *inverted* ovule. This is shown in Fig. 267, and 273 ; also a much enlarged section lengthwise, or diagram, in Fig. 275. To understand it, we have only to suppose the first sort (Fig. 270) to be inverted on its stalk, or rather to have its stalk bent round, applied to one side of the ovule lengthwise, and to grow fast to the coat down to near the orifice (f) ; the hilum, therefore, where the seed-stalk is to break away (h), is close to the orifice ; but the chalaza (c) is here at the top of the ovule ; between it and the hilum runs a ridge or cord, called the *Rhaphe* (r), which is simply that part of the stalk which, as the ovule grew and turned over, adhered to its surface. — Lastly, the

Amphitropous or *half-anatropous* ovule (Fig. 272) differs from the last only in having a shorter rhaphe, ending about half-way between the chalaza and the orifice. So the hilum or attachment is not far from the middle of one side, while the chalaza is at one end and the orifice at the other.

326. The internal structure of the ovule is sufficiently displayed in the subjoined diagrams, representing a longitudinal slice of two

FIG. 270. Orthotropous ovule of Buckwheat : *c*, hilum and chalaza ; *f*, orifice.
FIG. 271. Campylotropous ovule of a Chickweed : *c*, hilum and chalaza ; *f*, orifice.
FIG. 272. Amphitropous ovule of Mallow : *f*, orifice ; *h*, hilum ; *r*, rhaphe ; *c*, chalaza.
FIG. 273. Anatropous ovule of a Violet ; the parts lettered as in the last.

ovules ; Fig. 274, an orthotropous, Fig. 275, an anatropous ovule.
The letters correspond in the two ; *c*, the chalaza ; *f*, the orifice ;
r, rhaphe (of which there is of course none in Fig. 274) ; *p*, the
outer coat, called *primine ;* *s*, inner coat, called *secundine ;* *n*, nucleus or kernel.

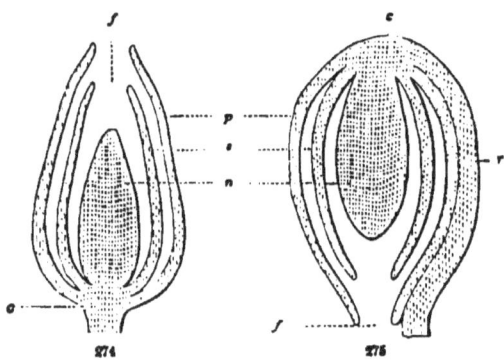

LESSON XIX.

MORPHOLOGY OF THE RECEPTACLE.

327. THE RECEPTACLE (also called the *Torus*) is the axis, or
stem, which the leaves and other parts of the blossom are attached
to (231). It is commonly small and short (as in Fig. 169) ; but it
sometimes occurs in more conspicuous and remarkable forms.

328. Occasionally it is elongated, as in some plants of the Caper
family (Fig. 276), making the flower really look like a branch, having its circles of leaves, stamens, &c., separated by long spaces or
internodes.

329. The Wild Geranium or Cranesbill has the receptacle prolonged above and between the insertion of the pistils, in the form
of a slender beak. In the blossom, and until the fruit is ripe, it
is concealed by the five pistils united around it, and their flat styles
covering its whole surface (Fig. 277). But at maturity, the five
small and one-seeded fruits separate, and so do their styles, from the
beak, and hang suspended from the summit. They split off elasti-

cally from the receptacle, curving upwards with a sudden jerk, which scatters the seed, often throwing it to a considerable distance.

330. When a flower bears a great many pistils, its receptacle is generally enlarged so as to give them room; sometimes becoming broad and flat, as in the Flowering Raspberry, sometimes elongated, as in the Blackberry, the Magnolia, &c. It is the receptacle in the Strawberry (Fig. 279), much

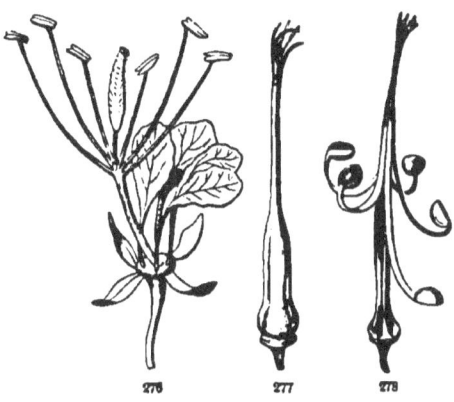

enlarged and pulpy when ripe, which forms the eatable part of the

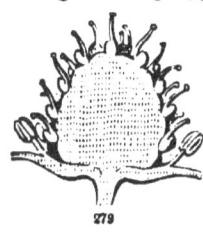

fruit, and bears the small seed-like pistils on its surface. In the Rose (Fig. 280), instead of being convex or conical, the receptacle is deeply concave, or urn-shaped. Indeed, a Rose-hip may be likened to a strawberry turned inside out, like the finger of a glove reversed, and the whole covered by the adherent tube of the calyx, which remains beneath in the strawberry.

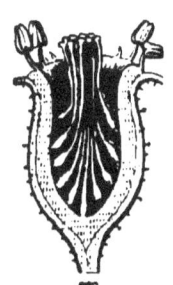

331. A **Disk** is a part of the receptacle, or a growth from it, enlarged under or around the pistil. It is *hypogynous* (269), when free from all union either with the pistil or the calyx, as in the Rue and the Orange (Fig. 281). It is *perigynous* (270), when it adheres to the base of the calyx, as in the Bladder-nut and Buckthorn (Fig. 282,

FIG. 276. Flower of Gynandropsis, the receptacle enlarged and flattened where it bears the sepals and petals, then elongated into a slender stalk, bearing the stamens (in appearance, but they are monadelphous) above its middle, and a compound ovary on its summit.

FIG. 277. Young fruit of the common Wild Cranesbill.

FIG. 278. The same, ripe, with the five pistils splitting away from the long beak or receptacle, and hanging from its top by their styles.

FIG. 279. Longitudinal section of a young strawberry, enlarged.

FIG. 280. Similar section of a young Rose-hip

FIG. 281. Pistil of the Orange, with a large hypogynous disk at its base.

283). Often it adheres both to the calyx and to the ovary, as in New Jersey Tea, the Apple, &c., consolidating the whole together. In such cases it is sometimes carried up and expanded on the top of

the ovary, as in the Parsley and the Ginseng families, when it is said to be *epigynous* (273).

332. In Nelumbium, — a large Water-Lily, abounding in the waters of our Western States, — the singular and greatly enlarged receptacle is shaped like a top, and bears the small pistils immersed in separate cavities of its flat upper surface (Fig. 284).

284

LESSON XX.

THE FRUIT.

333 THE ripened ovary, with its contents, becomes the *Fruit.* When the tube of the calyx adheres to the ovary, it also becomes a part of the fruit: sometimes it even forms the principal bulk of it, as in the apple and pear.

334. Some fruits, as they are commonly called, are not fruits at all in the strict botanical sense. A strawberry, for example (as we have just seen, 330, Fig. 282), although one of the choicest *fruits* in the common acceptation, is only an enlarged and pulpy receptacle, bearing the real fruits (that is, the ripened pistils) scattered over its

FIG. 282. Flower of a Buckthorn, with a large perigynous disk. 283. The same, divided. FIG. 284. Receptacle of Nelumbium, in fruit.

surface, and too small to be much noticed. And mulberries, figs, and pine-apples are masses of many fruits with a pulpy flower-stalk, &c. Passing these by for the present, let us now consider only

335. **Simple Fruits.** These are such as are formed by the ripening of a single pistil, whether simple (305) or compound (311).

336. A simple fruit consists, then, of the *Seed-vessel* (technically called the *Pericarp*), or the walls of the ovary matured, and the seeds contained in it. Its structure is generally the same as that of the ovary, but not always; because certain changes may take place after flowering. The commonest change is the obliteration in the growing fruit of some parts which existed in the pistil at the time of flowering. The ovary of a Horsechestnut, for instance, has three cells and two ovules in each cell; but the fruit never has more than three seeds, and rarely more than one or two, and only as many cells. Yet the vestiges of the seeds that have not matured, and of the wanting cells of the pod, may always be detected in the ripe fruit. This obliteration is more complete in the Oak and Chestnut. The ovary of the first likewise has three cells, that of the second six or seven cells, each with two ovules hanging from the summit. We might therefore expect the acorn and the chestnut to have as many cells, and two seeds in each cell. Whereas, in fact, all the cells and all the ovules but one are uniformly obliterated in the forming fruit, which thus becomes one-celled and one-seeded, and rarely can any vestige be found of the missing parts.

337. On the other hand, a one-celled ovary sometimes becomes several-celled in the fruit by the formation of false partitions, commonly by cross-partitions, as in the jointed pod of the Sea-Rocket and the Tick-Trefoil (Fig. 304).

338. **Their Kinds.** In defining the principal kinds of simple fruits which have particular names, we may classify them, in the first place, into, — 1. *Fleshy Fruits*; 2. *Stone Fruits*; and 3. *Dry Fruits*. The first and second are of course *indehiscent*; that is, they do not split open when ripe to discharge the seeds.

339. In *fleshy fruits* the whole pericarp, or wall of the ovary, thickens and becomes soft (fleshy, juicy, or pulpy) as it ripens. Of this the leading kind is

340. **The Berry,** such as the gooseberry and currant, the blueberry and cranberry, the tomato, and the grape. Here the whole flesh is equally soft throughout. The orange is merely a berry with a leathery rind.

341. **The Pepo,** or *Gourd-fruit,* is the sort of berry which belongs to the Gourd family, mostly with a hard rind and the inner portion softer. The pumpkin, squash, cucumber, and melon are the principal examples.

342. **The Pome** is a name applied to the apple, pear, and quince; fleshy fruits like a berry, but the principal thickness is calyx, only the papery pods arranged like a star in the core really belonging to the pistil itself (333).

343. Secondly, as to fruits which are partly fleshy and partly hard, one of the most familiar kinds is

344. **The Drupe,** or *Stone-fruit;* of which the cherry, plum, and 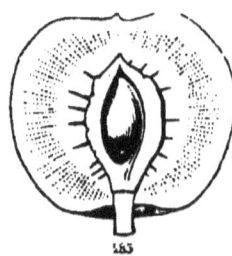 peach (Fig. 285) are familiar examples. In this the outer part of the thickness of the pericarp becomes fleshy, or softens, like a berry, while the inner hardens, like a nut. From the way in which the pistil is constructed (305), it is evident that the fleshy part here answers to the lower, and the stone to the upper, side of the leaf; — a leaf always consisting of two layers of green pulp, an upper and an under layer, which are considerably different (439).

345. Whenever the walls of a fruit are separable into two layers, the outer layer is called the *Exocarp,* the inner, the *Endocarp* (from Greek words meaning "outside fruit" and "inside fruit"). But in a drupe the outer portion, being fleshy, is likewise called *Sarcocarp* (which means "fleshy fruit"), and the inner, the *Putamen* or stone. The stone of a peach, and the like, it will be perceived, belongs to the fruit, not to the seed. When the walls are separable into three layers, the outer layer is named either exocarp or *Epicarp;* the middle one is called the *Mesocarp* (i. e. middle fruit); and the innermost, as before, the *Endocarp.*

346. Thirdly, in *dry fruits* the seed-vessel remains herbaceous in texture, or becomes thin and membranaceous, or else it hardens throughout. Some forms remain closed, that is, are *indehiscent* (338); others are *dehiscent,* that is, split open at maturity in some regular way. Of indehiscent or closed dry fruits the principal kinds are the following.

347. **The Achenium,** or *Akene,* is a small, one-seeded, dry, indehis-

cent fruit, such as is popularly taken for a naked seed : but it is plainly a ripened ovary, and shows the remains of its style or stigma, or the place from which it has fallen. Of this sort are the fruits of the Buttercup (Fig. 286, 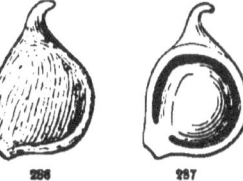 287), the Cinque-foil, and the Strawberry (Fig. 279, 288); that is, the real fruits, botanically speaking, of the latter, which are taken for seeds, not the large juicy receptacle on the surface of which they rest (339). Here the akenes are simple pistils (305), very numerous in the same flower, and forming a head of such fruits. In the Nettle, Hemp, &c., there is only one pistil to each blossom.

348. In the raspberry and blackberry, each grain is a similar pistil, like that of the strawberry in the flower, but ripening into a miniature stone-fruit, or drupe. So that in the strawberry we eat the receptacle, or end of the flower-stalk; in the raspberry, a cluster of stone-fruits, like cherries on a very small scale; and in the blackberry, both a juicy receptacle and a cluster of stone-fruits covering it (Fig. 289, 290).

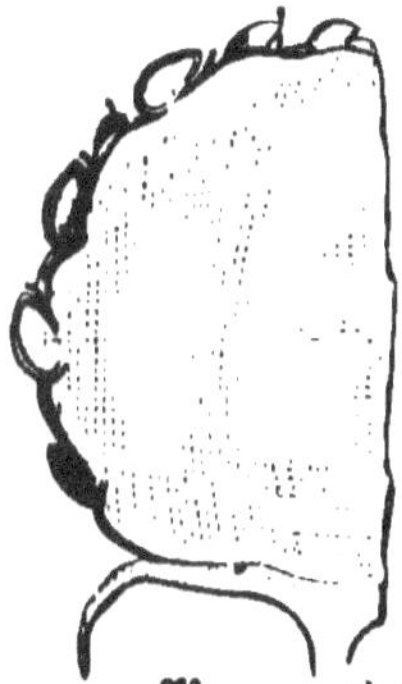

349. The fruit of the Composite family is also an achenium. Here the surface of the ovary is covered by an adherent calyx-tube, as is evident from the position of the corolla, apparently standing on its summit (321, and Fig. 220, a). Sometimes the limb or divisions of the calyx are entirely wanting, as in Mayweed (Fig. 291) and Whiteweed. Sometimes the limb of the calyx forms a *crown* or cup on the top of the achenium, as in Succory (Fig. 292); in Coreopsis, it often takes the form of two blunt teeth or scales; in the Sunflower (Fig. 293), it consists of two

FIG. 286. Achenium of Buttercup. 287. Same, cut through, to show the seed within.
FIG. 288. Slice of a part of a ripe strawberry, enlarged; some of the achenia shown cut through.
FIG. 289. Slice of a part of a blackberry. 290 One of the grains or drupes divided, more enlarged; showing the flesh, the stone, and the seed, as in Fig. 285.

thin scales which fall off at the touch; in the Sneezeweed, of about
five very thin scales, which look more like a calyx (Fig. 294); and
in the Thistle, Aster, Sow-Thistle (Fig. 295), and hundreds of others,
it is cut up into a tuft of fine bristles or hairs. This is called the
Pappus; — a name which properly means the down like that of the
Thistle; but it is applied to all these forms,
and to every other under which the limb of the
calyx of the " compound flowers " appears. In
Lettuce, Dandelion (Fig. 296), and the like,
the achenium as it matures tapers upwards
into a slender beak, like a stalk to the pappus.

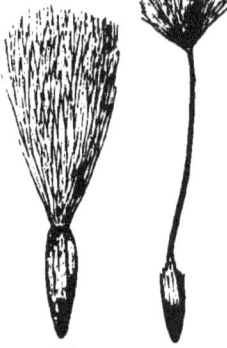

291   292   293   294   295   296

350. **A Utricle** is the same as an achenium, but with a thin and
bladdery loose pericarp; like that of the Goosefoot or Pigweed

297

(Fig. 297). When ripe it bursts open irregularly to
discharge the seed; or sometimes it opens by a circular
line all round, the upper part falling off like a lid; as in
the Amaranth (Fig. 298).

351. **A Caryopsis, or Grain,** differs from the last only
in the seed adhering to the thin pericarp
throughout, so that fruit and seed are in-
corporated into one body; as in wheat, Indian corn, and other kinds of grain.

298

352. **A Nut** is a dry and indehiscent fruit,
commonly one-celled and one-seeded, with a hard, crustaceous. or bony wall, such as the cocoanut, hazelnut,
chestnut, and the acorn (Fig. 21, 299). Here the
involucre, in the form of a cup at the base, is called the *Cupule.*   In
the Chestnut it forms the bur; in the Hazel, a leafy husk.

299

FIG. 291. Achenium of Mayweed (no pappus). 292. That of Succory (its pappus a shallow cup). 293. Of Sunflower (pappus of two deciduous scales). 294. Of Sneezeweed (Hele-
nium), with its pappus of five scales. 295. Of Sow-Thistle, with its pappus of delicate downy
hairs. 296. Of the Dandelion, its pappus raised on a long beak.

IG. 297. Utricle of the common Pigweed (Chenopodium album).

FIG. 298. Utricle (pyxis) of Amaranth, opening all round (circumcissile).

FIG. 299. Nut (acorn) of the Oak, with its cup (or cupule).

353. **A Samara, or Key-fruit,** is either a nut or an achenium, or any other indehiscent fruit, furnished with a wing, like that of the Maple (Fig. 1), Ash (Fig. 300), and Elm (Fig. 301).

354. **The Capsule, or Pod,** is the general name for dry seed-vessels which split or burst open at maturity. But several sorts of pod are distinguished by particular names. Two of them belong to simple pistils, namely, the *Follicle* and the *Legume*.

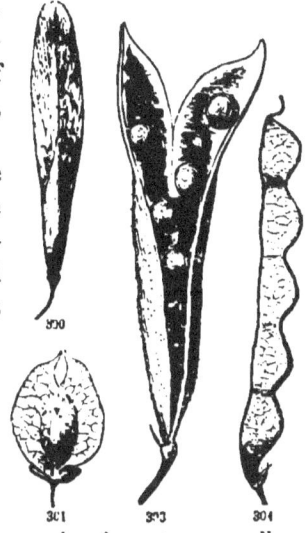

355. **The Follicle** is a fruit of a simple pistil opening along the inner suture (307). The pods of the Pæony, Columbine, Larkspur, Marsh-Marigold (Fig. 302), and Milkweed are of this kind. The seam along which the follicle opens answers to the edges of the pistil-leaf (Fig. 251, 253).

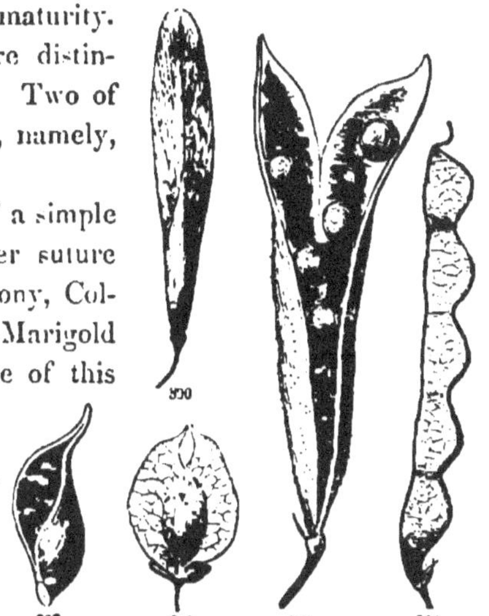

356. **The Legume** or true *Pod*, like the Pea-pod (Fig. 303), is similar to the follicle, only it opens by the outer as well as the inner or ventral suture (307), that is, by what answers to the midrib as well as by what answers to the united margins of the leaf. It splits therefore into two pieces, which are called *valves*. The legume belongs to plants of the Pulse family, which are accordingly termed *Leguminosæ*, that is, leguminous plants. So the fruits of this family keep the name of legume, whatever their form, and whether they open or not. A legume divided across into one-seeded joints, which separate when ripe, as in Tick-Trefoil (Fig. 304), is named a *Loment*.

357. **The true Capsule** is the pod of a compound pistil. Like the ovary it resulted from, it may be one-celled, or it may have as many cells as there are carpels in its composition. It may discharge its seeds through chinks or pores, as in the Poppy, or burst irregularly in some part, as in Lobelia and the Snapdragon ; but commonly it splits open (or is *dehiscent*) lengthwise into regular pieces, called *valves*.

FIG. 300. Samara or key of the White Ash.   301. Samara of the American Elm.
FIG. 302. Follicle of Marsh-Marigold (Caltha palustris).
FIG. 303. Legume of a Sweet Pea, opened.
FIG. 304. Loment or jointed legume of Tick-Trefoil (Desmodium).

358. *Dehiscence* of a pod resulting from a compound pistil, when regular, takes place in one of two principal ways, which are best shown in pods of two or three cells. Either the pod 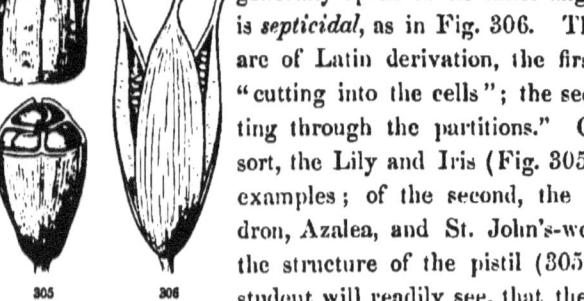 splits open down the middle of the back of each cell, when the dehiscence is *loculicidal*, as in Fig. 305 ; or it splits through the partitions, after which each cell generally opens at its inner angle, when it is *septicidal*, as in Fig. 306. These names are of Latin derivation, the first meaning "cutting into the cells"; the second, "cutting through the partitions." Of the first sort, the Lily and Iris (Fig. 305) are good examples ; of the second, the Rhododendron, Azalea, and St. John's-wort. From the structure of the pistil (305–311) the student will readily see, that the line down the back of each cell answers to the dorsal suture of the carpel ; so that the pod opens by this when loculicidal, while it separates into its component carpels, which open as follicles, when septicidal. Some pods open both ways, and so split into twice as many valves as the carpels of which they are formed.

359. In loculicidal dehiscence the valves naturally bear the partitions on their middle ; in the septicidal, half the thickness of a partition is borne on the margin of each valve. See the diagrams, Fig. 307 – 309. A variation of either mode sometimes occurs, as

shown in the diagram, Fig. 309, where the valves break away from the partitions. This is called *septifragal* dehiscence ; and may be seen in the Morning-Glory.

360. Three remaining sorts of pods are distinguished by proper names, viz. : —

FIG. 305. Capsule of Iris (with loculicidal dehiscence), below cut across.
FIG. 306. Pod of a Marsh St. John's-wort, with septicidal dehiscence.
FIG. 307. Diagram of septicidal ; 308, of loculicidal ; and 309, of septifragal dehiscence.

361. **The Silique** (Fig. 310), the peculiar pod of the Mustard family; which is two-celled by a false partition stretched across between two parietal placentæ. It generally opens by two valves from below upwards, and the placentæ with the partition are left behind when the valves fall off.

362. **A Silicle or Pouch** is only a short and broad silique, like that of the Shepherd's Purse, of the Candy-tuft, &c.

363. **The Pyxis** is a pod which opens by a circular horizontal line, the upper part forming a lid, as in Purslane (Fig. 311), the Plantain, Henbane, &c. In these the dehiscence extends all round, or is *circumcissile*. So it does in Fig. 298, which represents a sort of one-seeded pyxis. In Jeffersonia or Twin-leaf, the line does not separate quite round, but leaves a portion to form a hinge to the lid.

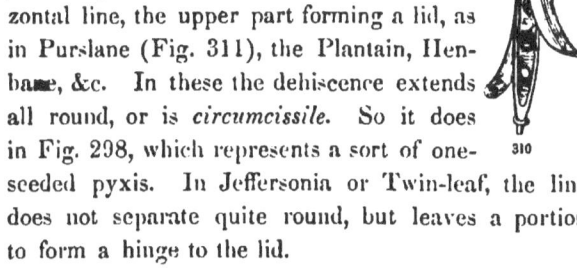

310

311

364. **Multiple or Collective Fruits** (334) are, properly speaking, masses of fruits, resulting from several or many blossoms, aggregated into one body. The pine-apple, mulberry, Osage-orange, and the fig. are fruits of this kind. This latter is a peculiar form, however, being to a mulberry nearly what a Rose-hip is to a strawberry (Fig. 279, 280), namely, with a hollow receptacle bearing the flowers concealed inside; and the whole eatable part is this pulpy common receptacle, or hollow thickened flower-stalk.

365. **A Strobile, or Cone** (Fig. 314), is the peculiar multiple fruit of Pines, Cypresses, and the like; hence named *Coniferæ*, viz. cone-bearing plants. As already shown (322), these cones are made of *open pistils*, mostly in the form of flat scales, regularly overlying each other, and pressed together in a spike or head. Each scale bears one or two naked seeds on its inner face. When the cone is ripe and dry, the scales turn back or diverge, and the seed peels off and falls, generally carrying with it a wing, which was a part of the lining of the scale, and which facilitates the dispersion of the seeds by the wind (Fig. 312, 313). In Arbor-Vitæ, the scales

312   313

FIG. 310.   Silique of Spring Cress (Cardamine rhomboidea), opening.
FIG. 311.   The pyxis, or pod, of the common Purslane
FIG. 312.   Inside view of a scale from the cone of Pitch-Pine; with one of the seeds (Fig. 313) detached; the other in its place on the scale.

of the small cone are few, and not very unlike the leaves (Fig. 265).
In Cypress they are very thick at the top and narrow at the base, so
as to make a peculiar sort of closed cone.   In Juniper and Red Cedar, the few scales of the very small cone become fleshy, and ripen
into a fruit which might be taken for a berry.

314

LESSON XXI.

THE SEED.

366. THE ovules (323), when they have an embryo (or undeveloped plantlet, 16) formed in them, become seeds.

367. The *Seed*, like the ovule from which it originates, consists
of its coats, or integuments, and a kernel.

368. The Seed-coats are commonly two (324), the outer and the
inner.   Fig. 315 shows the two, in a seed cut through
lengthwise.   The outer coat is often hard or crustaceous,
whence it is called the *Testa*, or shell of the seed; the
inner is thin and delicate.

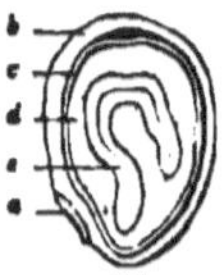

369. The shape and the markings, so various in different seeds, depend mostly on the outer coat.   Sometimes it fits

FIG. 314.   Cone of Pitch-Pine (Pinus rigida).
FIG. 315.   Seed of Basswood cut through lengthwise: *a*, the hilum or scar; *b*, the outer
coat; *c*, the inner; *d*, the albumen; *e*. the embryo.

the kernel closely; sometimes it is expanded into a *wing*, as in the Trumpet-Creeper (Fig. 316), and occasionally this wing is cut up into shreds or tufts, as in the Catalpa; or instead of a wing it may bear a *coma*, or tuft of long and soft hairs, such as we find in the Milkweed or Silkweed (Fig. 317). The object of wings or downy tufts is to render the seeds buoyant, so that they may be widely dispersed by the winds. This is clear, not only from their evident adaptation to this purpose, but also from the interesting fact

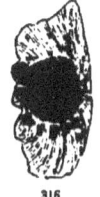

316

that winged and tufted seeds are found only in fruits that split open at maturity, never in those that remain closed. The coat of some seeds is beset with long hairs or wool. *Cotton*, one of the most important vegetable products, — since it forms the principal clothing of the larger part of the human race, — consists of the long and woolly hairs which thickly cover the whole surface of the seed. Certain seeds have an additional, but more or less incomplete covering, outside of the real seed-coats, called an

370. **Aril, cr Arillus.** The loose and transparent bag which encloses the seed of the White Water-Lily (Fig. 318) is of this kind. So is the *mace* of the nutmeg; and also the scarlet pulp around the seeds of the Waxwork (Celastrus) and Strawberry-bush (Euonymus), so ornamental in autumn, after the pods burst. The aril is a growth from the extremity of the seed-stalk, or the placenta.

371. The names of the parts of the seed and of its kinds are the same as in the ovule. The scar left where the seed-stalk separates is called the *Hilum*. The orifice of the ovule, now closed up, and showing only a small point or mark, is named the *Micropyle*. The terms *orthotropous, anatropous,* &c.

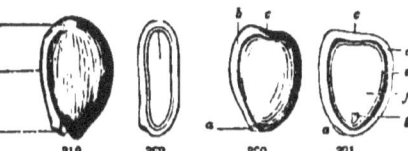

319   322   320   321

FIG. 316. A winged seed of the Trumpet-Creeper.

FIG. 317. Seed of Milkweed, with a *coma* or tuft of long silky hairs at one end.

FIG. 318. Seed of White Water-Lily, enclosed in its aril.

FIG. 319. Seed of a Violet (anatropous): *a*, hilum; *b*, rhaphe; *c*, chalaza.

FIG. 320. Seed of a Larkspur (also anatropous); the parts lettered as in the last.

FIG. 321. The same, cut through lengthwise: *a*, the hilum; *c*, chalaza; *d*, outer seed-coat; *e*, inner seed-coat; *f*, the albumen; *g*, the minute embryo.

FIG. 322. Seed of a St. John's-wort, divided lengthwise; here the whole kernel is embryo.

apply to seeds just as they do to ovules (325) ; and so do those
terms which express the direction of the ovule or the seed in the
cell ; such as *erect, ascending, horizontal, pendulous,* or *suspended*
(323) : therefore it is not necessary to explain them anew. The
accompanying figures (Fig. 319 – 322) show all the parts of the
most common kind of seed, namely, the anatropous.

372. **The Kernel, or Nucleus,** is the whole body of the seed within the
coats. In many seeds the kernel is all *Embryo ;* in others a large
part of it is the *Albumen.*

373. **The Albumen** of the seed is an accumulation of nourishing
matter (starch, &c.), commonly surrounding the embryo, and destined to nourish it when it begins to grow, as was explained in the
earlier Lessons (30 – 32). It is the floury part of wheat, corn (Fig.
38, 39), buckwheat, and the like. But it is not always *mealy* in
texture. In Poppy-seeds it is *oily.* In the seeds of Pæony and
Barberry, and in the cocoanut, it is *fleshy ;* in coffee it is *corneous*
(that is, hard and tough, like horn) ; in the Ivory Palm it has the
hardness as well as the general appearance of ivory, and is now
largely used as a substitute for it in the fabrication of small objects.
However solid its texture, the albumen always softens and partly
liquefies during germination ; when a considerable portion of it is
transformed into sugar, or into other forms of fluid nourishment, on
which the growing embryo may feed.

374. **The Embryo,** or *Germ,* is the part to which all the rest of the
seed, and also the fruit and the flower, are subservient. When the
embryo is small and its parts little developed, the albumen is the
more abundant, and makes up the principal bulk of the seed, as in
Fig. 30, 321, 325. On the other hand, in many seeds there is no
albumen at all ; but the strong embryo forms the whole kernel ; as
in the Maple (Fig. 2, 3), Pumpkin (Fig. 9), Almond, Plum, and
Apple (Fig. 11, 12), Beech (Fig. 13), and the like. Then, whatever nourishment is needed to establish the plantlet in the soil is
stored up in the body of the embryo itself, mostly in its seed-leaves.
And these accordingly often become very large and thick, as in the
almond, bean, and pea (Fig. 16, 19), acorn (Fig. 21), chestnut, and
horsechestnut (Fig. 23, 24). Besides these, Fig. 25, 26, 30 to 37,
43, and 45 exhibit various common forms of the embryo ; and also
some of the ways in which it is placed in the albumen ; being
sometimes straight, and sometimes variously coiled up or packed
away.

375. The embryo, being a rudimentary plantlet, ready formed in the seed, has only to grow and develop its parts to become a young plant (15). Even in the seed these parts are generally distinguishable, and are sometimes very conspicuous; as in a Pumpkin-seed, for example (Fig. 323, 324). They are, first,

376. The **Radicle**, or rudimentary stemlet, which is sometimes long and slender, and sometimes very short, as we may see in the numerous figures already referred to. In the seed it always points to the micropyle (371), or what answers to the foramen of the ovule (Fig. 325, 326). As to its position in the fruit, it is said to be *inferior* when it points to the base of the pericarp, *superior* when it points to its summit, &c. The base or free end of the radicle gives rise to the root; the other extremity bears

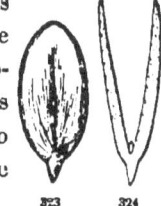

377. The **Cotyledons or Seed-Leaves.** With these in various forms we have already become familiar. The number of cotyledons has also been explained to be important (32, 33). In Corn (Fig. 40), and in all Grasses, Lilies, and the like, we have a *Monocotyledonous* embryo, namely, one furnished with only a single cotyledon or seed-leaf. — Nearly all the rest of our illustrations exhibit various forms of the *Dicotyledonous* embryo; namely, with a pair of cotyledons or seed-leaves, always opposite each other. In the Pine family we find a *Polycotyledonous* embryo (Fig. 45, 46); that is, one with several, or more than two, seed-leaves, arranged in a circle or whorl.

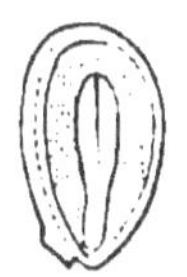

378. The **Plumule** is the little bud, or rudiment of the next leaf or pair of leaves after the seed-leaves. It appears at the summit of the radicle, between the cotyledons when there is a pair of them, as in Fig. 324, 14, 24, &c.; or the cotyledon when only one is wrapped round it, as in Indian Corn, Fig. 40. In germination the plumule develops upward, to form the ascending trunk or stem of the plant, while the other end of the radicle grows downward, and becomes the root.

FIG. 323. Embryo of the Pumpkin, seen flatwise. 324. Same cut through and viewed edgewise, enlarged; the small plumule seen between the cotyledons at their base.

FIG. 325. Seed of a Violet (Fig. 319) cut through, showing the embryo in the section, edgewise; being an *anatropous* seed, the radicle of the straight embryo points down to the base near the hilum.

FIG. 326. Similar section of the *orthotropous* seed of Buckwheat. Here the radicle points directly away from the hilum, and to the apex of the seed; also the thin cotyledons happen in this plant to be bent round into the same direction.

379. This completes the circle, and brings our vegetable history round to its starting-point in the Second Lesson ; namely, The Growth of the Plant from the Seed.

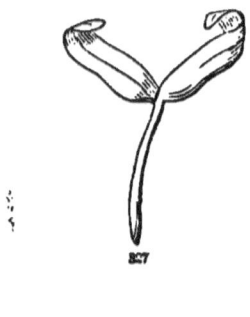

LESSON XXII.

HOW PLANTS GROW.

380. A PLANT grows from the seed, and from a tiny embryo, like that of the Maple (Fig. 327), becomes perhaps a large tree, producing every year a crop of seeds, to grow in their turn in the same way. But *how* does the plant grow? A little seedling, weighing only two or three grains, often doubles its weight every week of its early growth, and in time may develop into a huge bulk, of many tons' weight of vegetable matter. How is this done? What is vegetable matter? Where did it all come from? And by what means is it increased and accumulated in plants? Such questions as these will now naturally arise in any inquiring mind ; and we must try to answer them.

381. Growth *is the increase of a living thing in size and substance.* It appears so natural to us that plants and animals should grow, that people rarely think of it as requiring any explanation. They say that a thing is so because it grew so. Still we wish to know how the growth takes place.

382. Now, in the foregoing Lessons we explained the whole structure of the plant, with all its organs, by beginning with the seedling plantlet, and following it onward in its development through the

FIG. 327. Germinating embryo of a Maple.

whole course of vegetation (12, &c.). So, in attempting to learn how this growth took place, it will be best to adopt the same plan, and to commence with the commencement, that is, with the first formation of a plant. This may seem not so easy, because we have to begin with parts too small to be seen without a good microscope, and requiring much skill to dissect and exhibit. But it is by no means difficult to describe them; and with the aid of a few figures we may hope to make the whole matter clear.

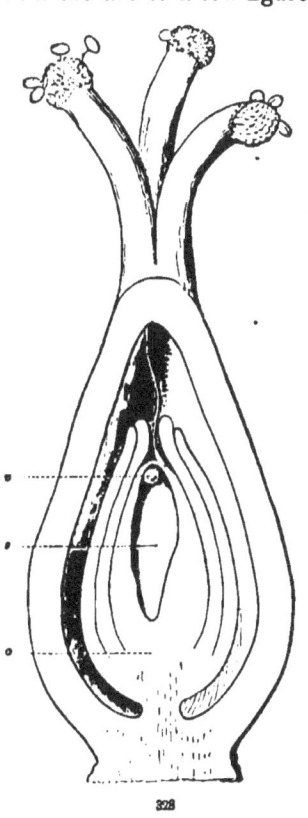

383. The embryo in the ripe seed is already a plant in miniature, as we have learned in the Second, Third, and Twenty-first Lessons. It is already provided with stem and leaves. To learn how the plant began, therefore, we must go back to an earlier period still; namely, to the formation and

384. **Growth of the Embryo** itself. For this purpose we return to the ovule in the pistil of the flower (323). During or soon after blossoming, a cavity appears in the kernel or nucleus of the ovule (Fig. 274, o), lined with a delicate membrane, and so forming a closed sac, named the *embryo-sac* (s). In this sac or cavity, at its upper end (viz. at the end next the orifice of the ovule), appears a roundish little *vesicle* or bladder-like body (v), perhaps less than one thousandth of an inch in diameter. This is the embryo, or rudimentary new plant, at its very beginning. But this vesicle never becomes anything more than a grain of soft pulp, unless the ovule has been acted upon by the pollen.

FIG. 328. Magnified pistil of Buckwheat; the ovary and ovule divided lengthwise: some pollen on the stigmas, one grain distinctly showing its tube, which penetrates the style, reappears in the cavity of the ovary, enters the mouth of the ovule (o), and reaches the surface of the embryo-sac (s), near the embryonal vesicle (v).

385. The pollen (297) which falls upon the stigma grows there in a peculiar way: its delicate inner coat extends into a tube (the pollen-tube), which sinks into the loose tissue of the stigma and the interior of the style, something as the root of a seedling sinks into the loose soil, reaches the cavity of the ovary, and at length penetrates the orifice of an ovule. The point of the pollen-tube reaches the surface of the embryo-sac, and in some unexplained way causes a particle of soft pulpy or mucilaginous matter (Fig. 328) to form a membranous coat and to expand into a vesicle, which is the germ of the embryo.

386. This vesicle (shown detached and more magnified in Fig. 329) is a specimen of what botanists call a *Cell.* Its wall of very delicate membrane encloses a mucilaginous liquid, in which there are often some minute grains, and commonly a larger soft mass (called its *nucleus*).

387. Growth takes place by this vesicle or cell, after enlarging to a certain size, dividing by the formation of a cross partition into two such cells, cohering together (Fig. 330); one of these into two more (Fig. 331); and these repeating the process by partitions formed in both directions (Fig. 332); forming a cluster or mass of cells, essentially like the first, and all proceeding from it. After increasing in number for some time in this way, and by a continuation of the same process, the embryo begins to shape itself; the upper end forms the radicle or root-end,

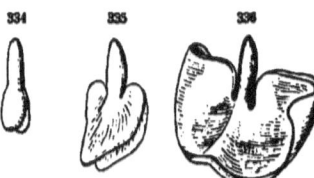

while the other end shows a notch between two lobes (Fig. 333), these lobes become the cotyledons or seed-leaves, and the embryo as it exists in the seed is at length completed (Fig. 336)

FIG. 329. Vesicle or first cell of the embryo, with a portion of the summit of the embryo-sac, detached. 330. Same, more advanced, divided into two cells. 331. Same, a little farther advanced, consisting of three cells. 332. Same, still more advanced, consisting of a little mass of young cells.

FIG. 333. Forming embryo of Buckwheat, moderately magnified, showing a nick at the end where the cotyledons are to be. 334. Same, more advanced in growth. 335. Same, still farther advanced. 336. The completed embryo, displayed and straightened out; the same as shown in a section when folded together in Fig. 326.

388. **The Growth of the Plantlet** when it springs from the seed is only a continuation of the same process. The bladder-like cells of which the embryo consists multiply in number by the repeated division of each cell into two. And the plantlet is merely the aggregation of a vastly larger number of these cells. This may be clearly ascertained by magnifying any part of a young plantlet. The young root, being more transparent than the rest, answers the purpose best. Fig. 56, on page 30, represents the end of the rootlet of Fig. 55, magnified enough to show the cells that form the surface. Fig. 337 and 338 are two small bits of the surface more highly magnified, showing the cells still larger. And if we make a thin slice through the young root both lengthwise and crosswise, and view it under a good microscope (Fig. 340), we may perceive that the whole interior is made up of just such cells. It is the same with the young stem and the leaves (Fig. 355, 357). It is essentially the same in the full-grown herb and the tree.

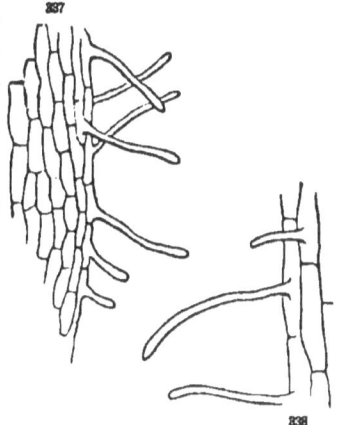

389. So the plant is an aggregation of countless millions of little vesicles, or cells (Fig. 339), as they are called, essentially like the cell it began with in the formation of the embryo (Fig. 329); and this first cell is the foundation of the whole structure, or the ancestor of all the rest. And a plant is a kind of structure built up of these individual cells, something as a house is built of bricks, — only the bricks or cells are not brought to the forming plant, but are made in it and by it; or, to give a better comparison, the plant is constructed much as a honeycomb is built up of cells, — only the plant constructs itself, and shapes its own materials into fitting forms.

390. And vegetable growth consists of two things ; — 1st, the expansion of each cell until it gets its full size (which is commonly not more than $\frac{1}{100}$ of an inch in diameter) ; and 2d, the multiplication

FIG. 337. Tissue from the rootlet of a seedling Maple, magnified, showing root-hairs. 339. A small portion, more magnified.
FIG. 339. A regularly twelve-sided cell, like those of Fig. 340, detached.

of the cells in number. It is by the latter, of course, that the principal increase of plants in bulk takes place.

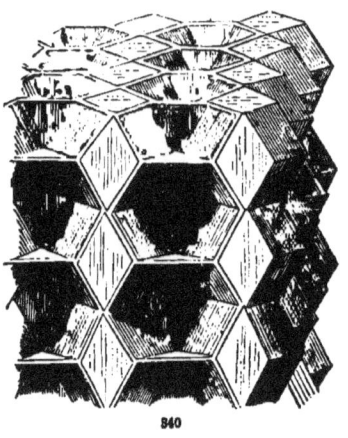

340

LESSON XXIII.

VEGETABLE FABRIC: CELLULAR TISSUE.

391. **Organic Structure.** A mineral — such as a crystal of spar, or a piece of marble — may be divided into smaller and still smaller pieces, and yet the minutest portion that can be seen with the microscope will have all the characters of the larger body, and be capable of still further subdivision, if we had the means of doing it, into just such particles, only of smaller size. A plant may also be divided into a number of similar parts : first into branches ; then each branch or stem, into joints or similar parts (34), each with its leaf or pair of leaves. But if we divide these into pieces, the pieces are not all alike, nor have they separately the. properties of the whole ; they are not whole things, but fragments or slices.

392. If now, under the microscope, we subdivide a leaf, or a piece of stem or root, we come down in the same way to the set of similar things it is made of, — to cavities with closed walls, — to *Cells*, as we call them (386), essentially the same everywhere, however they may vary in shape. These are the *units*, or the elements of which every part consists ; and it is their growth and their multiplication which

FIG. 340. Magnified view, or diagram, of some perfectly regular cellular tissue, formed of twelve-sided cells, cut crosswise and lengthwise.

make the growth of the plant, as was shown in the last Lesson. We cannot divide them into similar smaller parts having the properties of the whole, as we may any mineral body. We may cut them in pieces; but the pieces are only mutilated parts of a cell. This is a peculiarity of organic things (2, 3): it is *organic structure.* Being composed of cells, the main structure of plants is called

393. **Cellular Tissue.** The cells, as they multiply, build up the tissues or fabric of the plant, which, as we have said (389), may be likened to a wall or an edifice built of bricks, or still better to a honeycomb composed of ranges of cells (Fig. 340).

394. The walls of the cells are united where they touch each other; and so the partition appears to be a simple membrane, although it is really double; as may be shown by boiling the tissue a few minutes and then pulling the parts asunder. And in soft fruits the cells separate in ripening, although they were perfectly united into a tissue, when green, like that of Fig. 340.

395 In that figure the cells fit together perfectly, leaving no interstices, except a very small space at some of the corners. But in most leaves, the cells are loosely heaped together, leaving spaces or passages of all sizes (Fig. 336); and in the leaves and stems of aquatic and marsh plants, in particular, the cells are built up into narrow partitions, which form the sides of large and regular canals or passages (as shown in Fig. 341). These passages form the holes or cavities so conspicuous on cutting across any of these plants, and which are always filled with air. They may be likened to a stack of chimneys, built up of cells in place of bricks.

396. When small and irregular, the interstices are called *intercellular spaces* (that is, spaces between the cells). When large and regular, they are named *intercellular passages* or *air-passages.*

397. It will be noticed that in slices of the root, stem, or any tissue where the cells are not partly separate, the boundaries of the cells are usually more or less six-sided, like the cells of a honeycomb; and this is apt to be the case in whatever direction the slice is made, whether crosswise, lengthwise, or obliquely. The reason of this is easy to see. The natural figure of the cell is globular Cells which are not pressed upon by others are generally round or roundish (except when they grow in some particular direction), as we see in the green pulp of many leaves. When a quantity of spheres (such, for instance, as a pile of cannon-balls) are heaped up, each one in the interior of the heap is touched by twelve others. If the spheres be

soft and yielding, as young cells are, when pressed together they will become twelve-sided, like that in Fig. 339. And a section in any direction will be six-sided, as are the meshes in Fig. 340.

398. The size of the common cells of plants varies from about the thirtieth to the thousandth of an inch in diameter. An ordinary size is from $\frac{1}{300}$ to $\frac{1}{500}$ of an inch; so that there may generally be from 27 to 125 millions of cells in the compass of a cubic inch!

399. Now when it is remembered that many stems shoot up at the rate of an inch or two a day, and sometimes of three or four inches, knowing the size of the cells, we may form some conception of the rapidity of their formation. The giant Puff-ball has been known to enlarge from an inch or so to nearly a foot in diameter in a single night; but much of this is probably owing to expansion. We take therefore a more decisive, but equally extraordinary case, in the huge flowering stem of the Century-Plant. After waiting many years, or even for a century, to gather strength and materials for the effort, Century-Plants in our conservatories send up a flowering stalk, which grows day after day at the rate of a foot in twenty-four hours, and becomes about six inches in diameter. This, supposing the cells to average $\frac{1}{300}$ of an inch in diameter, requires the formation of over twenty thousand millions of cells in a day!

400. The walls of the cells are almost always colorless. The green color of leaves and young bark, and all the brilliant hues of flowers, are due to the contents of the cells, seen through their more or less transparent walls.

401. At first the walls are always very thin. In all soft parts they remain so; but in other cases they thicken on the inside and harden, as we see in the stone of stone-fruits, and in all hard wood (Fig. 345) Sometimes this thickening continues until the cell is nearly filled up solid.

402. The walls of cells are perfectly closed and whole, at least in all young and living cells. Those with thickened walls have thin places, indeed; but there are no holes opening from one cell into another. And yet through these closed cells the sap and all the juices are conveyed from one end of the plant to the other.

403. Vegetable cells may vary widely in shape, particularly when not combined into a tissue or solid fabric. The hairs of plants, for example, are cells drawn out into tubes, or are composed of a row of cells, growing on the surface. Cotton consists of simple long hairs on the coat of the seed; and these hairs are single cells. The hair-

like bodies which abound on young roots are very slender projections of some of the superficial cells, as is seen in Fig. 337. Even the fibres of wood, and what are called vessels in plants, are only peculiar forms or transformations of cells.

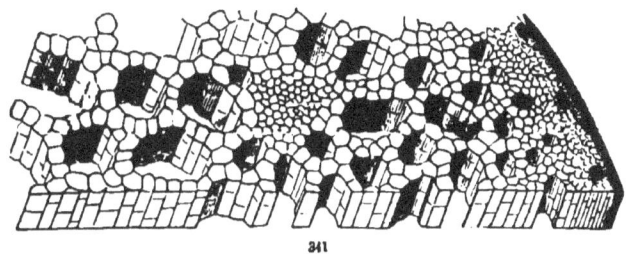

341

LESSON XXIV.

VEGETABLE FABRIC: WOOD.

404. CELLULAR TISSUE, such as described in the last Lesson, makes up the whole structure of all very young plants, and the whole of Mosses and other vegetables of the lowest grade, even when full grown. But this fabric is too tender or too brittle to give needful strength and toughness for plants which are to rise to any considerable height and support themselves. So all such plants have also in their composition more or less of

405. **Wood.** This is found in all common herbs, as well as in shrubs and trees; only there is not so much of it in proportion to the softer cellular tissue. It is formed very early in the growth of the root, stem, and leaves; traces of it appearing in large embryos even while yet in the seed.

406. Wood is likewise formed of cells, — of cells which at first are just like those that form the soft parts of plants. But early in their growth, some of these lengthen and at the same time thicken their walls; these are what is called *Woody Fibre* or *Wood-Cells ;* others grow to a greater size, have thin walls with various markings upon them, and often run together end to end so as to form pretty

large tubes, comparatively; these are called *Ducts*, or sometimes *Vessels.* Wood almost always consists of both woody fibres and ducts,

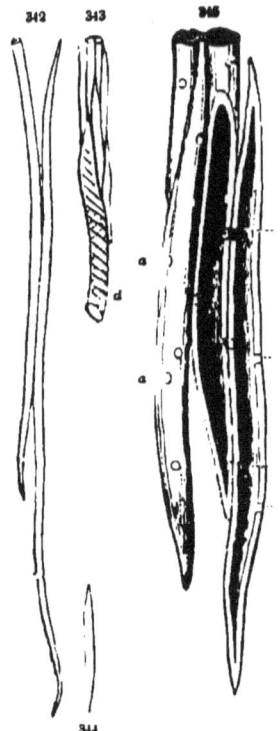

variously intermingled, and combined into bundles or threads which run lengthwise through the root and stem, and are spread out to form the framework of the leaves (136). In trees and shrubs they are so numerous and crowded together, that they make a solid mass of wood. In herbs they are fewer, and often scattered. That is all the difference.

407. The porosity of some kinds of wood, which is to be seen by the naked eye, as in mahogany and Oak-wood, is owing to a large sort of ducts. These generally contain air, except in very young parts, and in the spring of the year, when they are often gorged with sap, as we see in a wounded Grape-vine, or in the trunk of a Sugar-Maple at that time. But in woody plants through the season, the sap is usually carried up from the roots to the leaves by the

408. **Wood-Cells, or Woody Fibre.** (Fig. 342 – 345.) These are small tubes, commonly between one and two thousandths, but in Pine-wood sometimes two or three hundredths, of an inch in diameter. Those from the tough bark of the Basswood, shown in Fig. 342, are only the fifteen-hundredth of an inch wide. Those of Buttonwood (Fig. 345) are larger, and are here highly magnified besides. They also show the way wood-cells are commonly put together, namely, with their tapering ends overlapping each other, — spliced together, as it were, — thus giving more strength and toughness to the stem, &c.

FIG. 342. Two wood-cells from the inner or fibrous bark of the Linden or Basswood. 343. Some tissue of the wood of the same, viz. wood-cells, and below (*d*) a portion of a spirally marked duct. 344. A separate wood-cell. All equally magnified.

FIG. 345. Some wood-cells of Buttonwood, highly magnified: *a*, thin spots in the walls, looking like holes; on the right-hand side, where the walls are cut through, these (*b*) are seen in profile.

409. In hard woods, such as Hickory, Oak, and Buttonwood (Fig. 345), the walls of these tubes are very thick, as well as dense ; while in soft woods, such as White-Pine and Basswood, they are pretty thin.

410. Wood-cells, like other cells (at least when young and living), have no openings ; each has its own cavity, closed and independent. They do not form anything like a set of pipes opening one into another, so as to convey an unbroken stream of sap through the plant, in the way people generally suppose. The contents can pass from one cell to another only by getting through the partitions in some way or other. And so short are the individual wood-cells generally, that, to rise a foot in such a tree as the Basswood, the sap has to pass through about two thousand partitions !

411. But although there are no holes (except by breaking away when old), there are plenty of thin places, which look like perforations; and through these the sap is readily transferred from one cell to another, in a manner to be explained further on (487). Some of them

346　　　347

are exhibited in Fig. 345, both as looked directly down upon, when they appear as dots or holes, and in profile where the cells are cut through. The latter view shows what they really are, namely, very thin places in the thickness of the wall ; and also that a thin place in one cell exactly corresponds to one in the contiguous wall of the next cell. In the wood of the Pine family, these thin spots are much larger, and are very conspicuous in a thin slice of wood under the microscope (Fig. 346, 347) ; — forming stamps impressed as it were upon each fibre of every tree of this great family, by which it may be known even in the smallest fragment of its wood.

\ 412. Wood-cells in the bark are generally longer, finer, and tougher than those of the proper wood, and appear more like fibres. For example, Fig. 344 represents a cell of the wood of Basswood, of average length, and Fig. 342 one (and part of another) of the fibrous bark, both drawn to the same scale. As these long cells form the principal part of fibrous bark, or *bast*, they are named *Bast-cells* or *Bast-fibres*. These give the great toughness to the inner bark of Basswood (i. e. Bast-wood) and of Leatherwood ; and they

FIG. 346. A bit of Pine-shaving, highly magnified, showing the large circular thin spots of the wall of the wood-cells. 347. A separate wood-cell, more magnified, the varying thickness of the wall at these spots showing as rings.

furnish the invaluable fibres of flax and hemp; the wood of the stem being tender, brittle, and destroyed by the processes which separate for use the tough and slender bast-cells.

413. **Ducts** (Fig. 348 – 350) are larger than wood-cells, some of them having a calibre large enough to be seen by the naked eye,

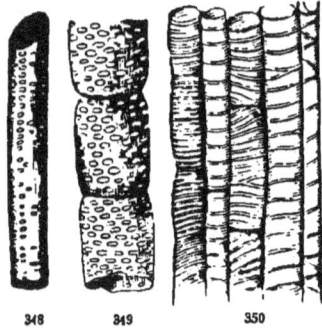

when cut across (407), although they are usually much too small for this. They are either long single cells, or are formed of a row of cells placed end to end. Fig. 349, a piece of a large dotted duct, and two of the ducts in Fig. 350, show this by their joints, which mark the boundaries of the several cells they are composed of.

348   349   350

414. The walls of ducts under the microscope display various kinds of markings. In what are called

Dotted Ducts (Fig. 348, 349), which are the commonest and the largest of all, — their cut ends making the visible porosity of Oak-wood, — the whole wall is apparently riddled with holes; but until they become old, these are only thin places.

Spiral Ducts, or *Spiral Vessels*, also the varieties of these called *Annular* or *Banded Ducts* (Fig. 350), are marked by a delicate fibre spirally coiled, or by rings or bands, thickening the wall. In the genuine spiral duct, the thread may be uncoiled, tearing the transparent wall in pieces; — as may be seen by breaking most young shoots, or the leaves of Strawberry or Amaryllis, and pulling the broken ends gently asunder, uncoiling these gossamer threads in abundance. In Fig. 355, some of these various sorts of ducts or vessels are shown in their place in the wood.

415. *Milk-Vessels*, *Turpentine-Vessels*, *Oil-Receptacles*, and the like, are generally canals or cavities formed between or among the cells, and filled with the particular products of the plant.

FIG. 348. Part of a dotted duct from a Grape-vine. 349. A similar one, evidently composed of a row of cells. 350. Part of a bundle of spiral and annular ducts from the stem of Polygonum orientale, or Princes' Feather. All highly magnified.

LESSON XXV.

ANATOMY OF THE ROOT, STEM, AND LEAVES.

416. Having in the last preceding Lessons learned what the materials of the vegetable fabric are, we may now briefly consider how they are put together, and how they act in carrying on the plant's operations.

417. The root and the stem are so much alike in their internal structure, that a description of the anatomy of the latter will answer for the former also.

418. **The Structure of the Rootlets,** however, or the tip of the root, demands a moment's attention. The tip of the root is the newest part, and is constantly renewing itself so long as the plant is active (67). It is shown magnified in Fig. 56, and is the same in all rootlets as in the first root of the seedling. The new roots, or their new parts, are mainly concerned in imbibing moisture from the ground; and the newer they are, the more actively do they absorb. The absorbing ends of roots are entirely composed of soft, new, and very thin-walled cellular tissue; it is only farther back that some wood-cells and ducts are found. The moisture (and probably also air) presented to them is absorbed through the delicate walls, which, like those of the cells in the interior, are destitute of openings or pores visible even under the highest possible magnifying power.

419. But as the rootlet grows older, the cells of its external layer harden their walls, and form a sort of skin, or *epidermis* (like that which everywhere covers the stem and foliage above ground), which greatly checks absorption. Roots accordingly cease very actively to imbibe moisture almost as soon as they stop growing (67).

420. Many of the cells of the surface of young rootlets send out a prolongation in the form of a slender hair-like tube, closed of course at the apex, but at the base opening into the cavity of the cell. These tubes or *root-hairs* (shown in Fig. 55 and 56, and a few of them, more magnified, in Fig. 337 and 338), sent out in all directions into the soil, vastly increase the amount of absorbing surface which the root presents to it.

421. **Structure of the Stem** (also of the body of the root). At the beginning, when the root and stem spring from the seed, they consist

13 *

almost entirely of soft and tender cellular tissue.   But as they grow, wood begins at once to be formed in them.

422. This woody material is arranged in the stem in two very different ways in different plants, making two sorts of wood. One sort we see in a Palm-stem, a rattan, and a Corn-stalk (Fig. 351); the other we are familiar with in Oak, Maple, and all our common kinds of wood.   In the first, the wood is made up of separate threads, scattered here and there throughout the whole diameter of the stem. In the second the wood is all collected to form a layer (in a slice across appearing as a ring) of wood, between a central cellular part which has none in it, the *Pith*, and an outer cellular part, the *Bark*. This last is the plan of all our Northern trees and shrubs, and of the greater part of our herbs.   The first kind is

423. **The Endogenous Stem**; so named from two Greek words meaning "inside-growing," because, when it lasts from year to year, the

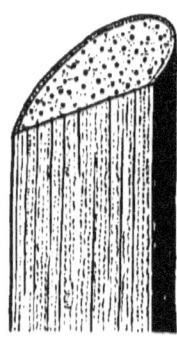

new wood which is added is interspersed among the older threads of wood, and in old stems the hardest and oldest wood is near the surface, and the youngest and softest towards the centre. All the plants represented in Fig. 47, on p. 19, (except the anomalous Cycas,) are examples of Endogenous stems.   And all such belong to plants with only one cotyledon or seed-leaf to the embryo (32).   Botanists therefore call them *Endogenous* or *Monocotyledonous Plants*, using sometimes one name, and sometimes the other.   Endogenous stems have no separate pith in the centre, no distinct bark, and no layer or ring of wood between these two; but the threads of wood are scattered throughout the whole, without any particular order. This is very different from

424. **The Exogenous Stem**, the one we have most to do with, since all our Northern trees and shrubs are constructed on this plan.   It belongs to all plants which have two cotyledons to the embryo (or more than two, such as Pines, 33); so that we call these either *Exogenous* or *Dicotyledonous Plants* (16), accordingly as we take the name from the stem or from the embryo.

425. In the Exogenous stem, as already stated, the wood is all collected into one zone, surrounding a pith of pure cellular tissue in the centre, and surrounded by a distinct and separable bark, the

FIG. 351. Section of a Corn-stalk (an endogenous stem), both crosswise and lengthwise.

outer part of which is also cellular. This structure is very familiar in common wood. It is really just the same in the stem of an herb, only the wood is much less in quantity. Compare, for

instance, a cross-section of the stem of Flax (Fig. 352) with that of a shoot of Maple or Horsechestnut of the same age. In an herb, the wood at the beginning consists of separate threads or little wedges of wood; but these, however few and scattered they may be, are all so placed in the stem as to mark out a zone (or in the cross-section a ring) of wood, dividing the pith within from the bark without.

426. The accompanying figures (which are diagrams rather than exact delineations) may serve to illustrate the anatomy of a woody exogenous stem, of one year old. The parts are explained in the references below. In the centre is the *Pith.* Surrounding this is the layer

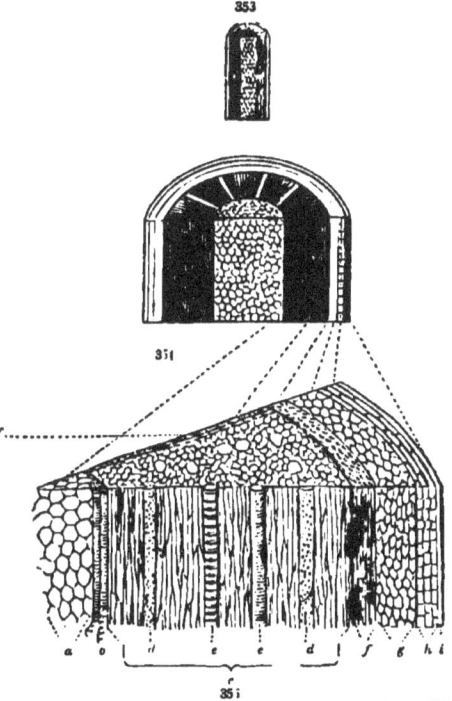

of *Wood*, consisting both of wood-cells and of ducts or vessels. From the pith to the bark on all sides run a set of narrow plates of cellular tissue, called *Medullary Rays:* these make the *silver-grain* of wood. On the cross-section they appear merely as narrow lines; but in wood cut lengthwise parallel to them, their faces show as glimmer-

FIG. 352. Cross-section of the stem of Flax, showing its bark, wood, and pith.

FIG. 353. Piece of a stem of Soft Maple, of a year old, cut crosswise and lengthwise.

FIG. 354. A portion of the same, magnified.

FIG. 355. A small piece of the same, taken from one side, reaching from the bark to the pith, and highly magnified : *a,* a small bit of the pith ; *b,* spiral ducts of what is called the *medullary sheath; c,* the wood ; *d, d,* dotted ducts in the wood ; *e, e,* annular ducts ; *f,* the liber or inner bark ; *g,* the green bark ; *h,* the corky layer ; *i,* the skin, or epidermis ; *f,* one of the medullary rays, or plates of silver-grain, seen on the cross-section.

ing plates, giving a peculiar appearance to Oak, Maple, and other wood with large medullary rays.

427. *The Bark* covers and protects the wood. At first it is all cellular, like the pith; but soon some slender woody fibres, called bast-cells (Fig. 342), generally appear in it, next the wood, forming

The Liber, or Fibrous Bark, the inner bark; to which belongs the fine fibrous *bast* or *bass* of Basswood, and the tough and slender fibres of flax and hemp, which are spun and woven, or made into cordage. In the Birch and Beech the inner bark has few if any bast-cells in its composition.

The Cellular or Outer Bark consists of cellular tissue only. It is distinguished into two parts, an inner and an outer, viz. : —

The Green Bark, or Green Layer, which consists of tender cells, containing the same green matter as the leaves, and serving the same purpose. In the course of the first season, in woody stems, this becomes covered with

The Corky Layer, so named because it is the same substance as *cork*; common cork being the thick corky layer of the bark of the Cork-Oak, of Spain. It is this which gives to the stems or twigs of shrubs and trees the aspect and the color peculiar to each; namely, light gray in the Ash, purple in the Red Maple, red in several Dog-woods, &c. Lastly,

The Epidermis, or skin of the plant, consisting of a layer of thick-sided empty cells, covers the whole.

428. **Growth of the Stem year after year.** So much for an exogenous stem only one year old. The stems of herbs perish at the end of the season. But those of shrubs and trees make a new growth every year. It is from their mode of growth in diameter that they take the name of *exogenous*, i. e. *outside-growing*. The second year, such a stem forms a second layer of wood outside of the first; the third year, another outside of that; and so on, as long as the tree lives. So that the trunk of an exogenous tree, when cut off at the base, exhibits as many concentric rings of wood as it is years old. Over twelve hundred layers have actually been counted on the stump of an aged tree, such as the Giant Cedar or Redwood of California; and there are doubtless some trees now standing in various parts of the world which were already in existence at the beginning of the Christian era.

429. As to the bark, the green layer seldom grows much after the first season. Sometimes the corky layer grows and forms new layers, inside of the old, for a good many years, as in the Cork-Oak,

the Sweet Gum-tree, and the White and the Paper Birch. But it all dies after a while; and the continual enlargement of the wood within finally stretches it more than it can bear, and sooner or later cracks and rends it, while the weather acts powerfully upon its surface; so the older bark perishes and falls away piecemeal year by year.

430. But the inner bark, or liber, does make a new growth annually, as long as the tree lives, inside of that formed the year before, and next the surface of the wood. More commonly the liber occurs in the form of thin layers, which may be distinctly counted, as in Basswood: but this is not always the case. After the outer bark is destroyed, the older and dead layers of the inner bark are also exposed to the weather, are riven or split into fragments, and fall away in succession. In many trees the bark acquires a considerable thickness on old trunks, although all except the innermost portion is dead; in others it falls off more rapidly; in the stems of Honeysuckles and Grape-vines, the bark all separates and hangs in loose shreds when only a year or two old.

431. Sap-wood. In the wood, on the contrary, — owing to its growing on the outside alone, — the older layers are quietly buried under the newer ones, and protected by them from all disturbance. All the wood of the young sapling may be alive, and all its cells or woody tubes active in carrying up the sap from the roots to the leaves. It is all *Sap-wood* or *Alburnum*, as young and fresh wood is called. But the older layers, removed a step farther every year from the region of growth, — or rather the zone of growth every year removed a step farther from them, — soon cease to bear much, if any, part in the circulation of the tree, and probably have long before ceased to be alive. Sooner or later, according to the kind of tree, they are turned into

432. Heart-wood, which we know is drier, harder, more solid, and much more durable as timber, than sap-wood. It is generally of a different color, and it exhibits in different species the hue peculiar to each, such as reddish in Red-Cedar, brown in Black-Walnut, black in Ebony, &c. The change of sap-wood into heart-wood results from the thickening of the walls of the wood-cells by the deposition of hard matter, lining the tubes and diminishing their calibre; and by the deposition of a vegetable coloring-matter peculiar to each species.

433. The heart-wood, being no longer a living part, may decay

S & F—8

and often does so, without the least injury to the tree, except by impairing the strength of the trunk, and so rendering it more liable to be overthrown.

434. **The Living Parts of a Tree,** of the exogenous kind, are only these: first, the rootlets at one extremity; second, the buds and leaves of the season at the other; and third, a zone consisting of the newest wood and the newest bark, connecting the rootlets with the buds or leaves, however widely separated these may be, — in the largest trees from two to four hundred feet apart. And these parts of the tree are all renewed every year. No wonder, therefore, that trees may live so long, since they annually reproduce everything that is essential to their life and growth, and since only a very small part of their bulk is alive at once. The tree survives, but nothing now living has existed long. In it, as elsewhere, life is a transitory thing, ever abandoning the *old*, and displaying itself afresh in the *new*.

435. **Cambium-Layer.** The new growth in the stem, by which it increases in diameter year after year, is confined to a narrow line between the wood and the inner bark. *Cambium* is the old name for the mucilage which is so abundant between the bark and the wood in spring. It was supposed to be poured out there, and that the bark really separated from the wood at this time. This is not the case. The newest bark and wood are still united by a delicate tissue of young and forming cells, — called the *Cambium-layer*, — loaded with a rich mucilaginous sap, and so tender that in spring the bark may be raised from the wood by the slightest force. Here, nourished by this rich mucilage, new cells are rapidly forming by division (387 – 390); the inner ones are added to the wood, and the outer to the bark, so producing the annual layers of the two, which are ever renewing the life of the trunk.

436. At the same time new rootlets, growing in a similar way, are extending the roots beneath; and new shoots, charged with new buds, annually develop fresh crops of leaves in the air above. Only, while the additions to the wood and bark remain as a permanent portion of the tree, or until destroyed by decay, the foliage is temporary, the crop of leaves being annually thrown off after they have served their purpose.

437. **Structure of the Leaf.** Leaves also consist both of a woody and a cellular part (135). The woody part is the framework of ribs and veins, which have already been described in full (136 – 147).

They serve not only to strengthen the leaf, but also to bring in the ascending sap, and to distribute it by the veinlets throughout every part. The cellular portion is the green pulp, and is nearly the same as the green layer of the bark. So that the leaf may properly enough be regarded as a sort of expansion of the fibrous and green layers of the bark. It has of course no corky layer ; but the whole is covered by a transparent skin or *epidermis*, resembling that of the stem.

438. The green pulp consists of cells of various forms, usually loosely arranged, so as to leave many irregular spaces, or air-passages, communicating with each other throughout the whole interior of the leaf (Fig. 356). The green color is owing to a peculiar green matter lying loose in the cells, in *form* of minute grains, named *Chlorophyll* (i. e. the green of leaves). It is this substance, seen through the transparent walls of the cells where it is accumulated, which gives the common green hue to vegetation, and especially to foliage.

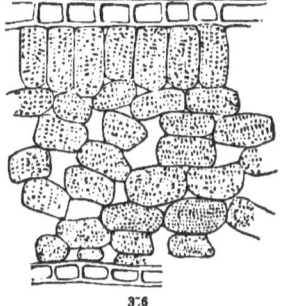

356

439. The green pulp in most leaves forms two principal layers ; an upper one, facing the sky, and an under one, facing the ground. The upper one is always deeper green in color than the lower. This is partly owing, perhaps, to a greater amount of chlorophyll in the upper cells, but mainly to the more compact arrangement of these cells. As is seen in Fig. 356 and 357, the cells of the upper side are oblong or cylindrical, and stand endwise to the surface of the leaf, usually close together, leaving hardly any vacant spaces. Those of the lower part of the leaf are apt to be irregular in shape, most of them with their longer diameter parallel to the face of the leaf, and are very loosely arranged, leaving many and wide air-chambers. The green color underneath is therefore diluted and paler.

440. In many plants which grow where they are subject to drought, and which hold their leaves during the dry season (the Oleander for example), the greater part of the thickness of the leaf consists of layers of long cells, placed endwise and very much com-

FIG. 356. Section through the thickness of a leaf of the Star Anise (Illicium), of Florida, magnified. The upper and the lower layers of thick-walled and empty cells represent the epidermis or skin. All those between are cells of the green pulp, containing grains of chlorophyll.

pacted, so as to expose as little surface as possible to the direct action of the hot sun. On the other hand, the leaves of marsh plants, and of others not intended to survive a drought, have their cells more loosely arranged throughout. In such leaves the epidermis, or skin, is made of only one layer of cells ; while in the Oleander, and the like, it consists of three or four layers of hard and thick-walled cells. In all this, therefore, we plainly see an arrangement for tempering the action of direct sunshine, and for restraining a too copious evaporation, which would dry up and destroy the tender cells, at least when moisture is not abundantly supplied through the roots.

441. That the upper side of the leaf alone is so constructed as to bear the sunshine, is shown by what happens when their position is reversed : then the leaf soon twists on its stalk, so as to turn again its under surface away from the light; and when prevented from doing so, it perishes.

442. A large part of the moisture which the roots of a growing plant are constantly absorbing, after being carried up through the stem, is evaporated from the leaves. A Sunflower-plant, a little over three feet high, and with between five and six thousand square inches of surface in foliage, &c., has been found to exhale twenty or thirty ounces (between one and two pints) of water in a day. Some part of this, no doubt, flies off through the walls of the epidermis or skin, at least in sunshine and dry weather; but no considerable portion of it. The very object of this skin is to restrain evaporation. The greater part of the moisture exhaled escapes from the leaf through the

443. **Stomates or Breathing-pores.** These are small openings through the epidermis into the air-chambers, establishing a direct communication between the whole interior of the leaf and the external air. Through these the vapor of water and air can freely escape, or enter, as the case may be. The aperture is guarded by a pair of thin-walled cells, — resembling those of the green pulp within, — which open when moist so as to allow exhalation to go on, but promptly close when dry, so as to arrest it before the interior of the leaf is injured by the dryness.

444. Like the air-chambers, the breathing-pores belong mainly to the under side of the leaf. In the White Lily, — where they are unusually large, and easily seen by a simple microscope of moderate power, — there are about 60,000 to the square inch on the epidermis of the lower surface of the leaf, and only about 3,000 in

the same space of the upper surface. More commonly there are few or none on the upper side ; direct sunshine evidently being unfavorable to their operation. Their immense numbers make up for their minuteness. They are said to vary from less than 1,000 to 170,000 to the square inch of surface. In the Apple-tree, where they are under the average as to number, there are about 24,000 to the square inch of the lower surface ; so that each leaf has not far from 100,000 of these openings or mouths.

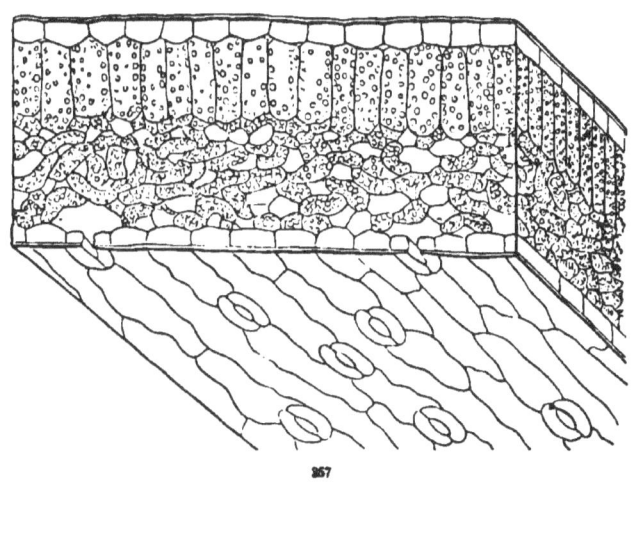

257

LESSON XXVI.

THE PLANT IN ACTION, DOING THE WORK OF VEGETATION.

445. BEING now acquainted with the machinery of the plant, we naturally proceed to inquire what the use of it is, and how it works.

446. It has already been stated, in the first of these Lessons (7), that the great work of plants is *to change inorganic into organic matter ;* that is, to take portions of earth and air, — of mineral matter, — upon which animals cannot live at all, and to convert them

FIG. 257. Portion of a White-Lily leaf, cut through and magnified, showing a section of the thickness, and also a part of the skin of the lower side, with some breathing-pores.

into something upon which they can live, namely, into food. All
the food of all animals is produced by plants. Animals live upon
vegetables; and vegetables live upon earth and air, principally
upon the air.

447. **Plants feed upon Earth and Air.** This is evident enough from
the way in which they live. Many plants will flourish in pure sand
or powdered chalk, or on the bare face of a rock or wall, watered
merely with rain-water. And almost any plant may be made to
grow from the seed in pure sand, and increase its weight many times.
even if it will not come to perfection. Many naturally live suspended
from the branches of trees high in the air, and nourished by it alone,
never having any connection with the soil (81); and some which
naturally grow on the ground, like the Live-for-ever of the gardens,
when pulled up by the roots and hung in the air will often flourish
the whole summer long.

448. It is true that fast-growing plants, or those which produce
considerable vegetable matter in one season, — especially in such a
concentrated form as to be useful as food for man or the higher
animals, — will come to maturity only in an enriched soil. But
what is a rich soil? One which contains decomposing vegetable
matter. or some decomposing animal matter; that is, in either case,
some decomposing organic matter formerly produced by plants;
aided by this, grain-bearing and other important vegetables will
grow more rapidly and vigorously, and make a greater amount of
nourishing matter, than they could if left to do the whole work at
once from the beginning. So that in these cases also all the organic
matter was made by plants, and made out of earth and air.

449. **Their Chemical Composition shows what Plants are made of.** The
soil and the air in which plants live, and by which they are everywhere surrounded, supply a variety of materials, some likely to be
useful to the plant, others not. To know what elements the plant
makes use of, we must first know of what its fabric and its products
are composed.

450. We may distinguish two sorts of materials in plants, one of
which is absolutely essential, and is the same in all of them; the
other, also to some extent essential, but very variable in different
plants, or in the same plant under different circumstances. The
former is the *organic*, the latter the *inorganic* or *earthy* materials.

451. **The Earthy or Inorganic Constituents.** If we burn thoroughly a
leaf, a piece of wood, or any other part of a vegetable, almost all of

it is dissipated into air.   But a little ashes remain : these represent
the earthy constituents of the plant.

452. They consist of some *potash* (or *soda* if a marine plant was
used), some *silex* (the same as flint), and probably a little *lime, al-
umine*, or *magnesia, iron* or *manganese, sulphur* or *phosphorus*, &c.
Some or all of these 'elements may be detected in many or most
plants.   But they make no part of their real fabric ; and they form
only from one or two to nine or ten parts out of a hundred of any
vegetable substance.   The ashes vary according to the nature
of the soil.   In fact, they consist, principally, of such materials as
happened to be dissolved, in small quantity, in the water which was
taken up by the roots ; and when that is consumed by the plant, or
flies off pure (as it largely does, 447) by exhalation, the earthy matter is left behind in the cells, — just as it is left incrusting the sides
of a teakettle in which much hard water has been boiled.   As is
very natural, therefore, we find more earthy matter (i. e. more
ashes) in the leaves than in any other part (sometimes as much as
seven per cent, when the wood contains only two per cent) ; because
it is through the leaves that most of the water escapes from the plant.
These earthy constituents are often useful to the plant (the silex, for
instance, increases the strength of the Wheat-stalk), or are useful in
the plant's products as furnishing needful elements in the food of man
and other animals ; and some must be held to be necessary to vegetation, since this is never known to go on without them.

453. **The Organic Constituents.**   As has just been remarked, when
we burn in the open air a piece of any plant, nearly its whole bulk,
and from 88 to more than 99 parts out of a hundred by weight of its
substance, disappear, being turned into air and vapor.   These are
the *organic constituents* which have thus been consumed, — the
actual materials of the cells and the whole real fabric of the plant.
And we may state that, in burning, it has been decomposed into exactly the same kinds of air, and the vapor of water, that the plant
used in its making.   The burning has merely undone the work of
vegetation, and given back the materials to the air just in the state
in which the plant took them.

454. It will not be difficult to understand what the organic constituents, that is, what the real materials, of the plant are, and how
the plant obtains them.   The substance of which vegetable tissue,
viz. the wall of the cells, is made, is by chemists named *Cellulose*.   It
is just the same thing in composition in wood and in soft cellular tis-

sue,—in the tender pot-herb and in the oldest tree. It is composed of carbon, hydrogen, and oxygen, 6 parts of the first to 10 of the second and 5 of the third. These, accordingly, are necessary materials of vegetable growth, and must be received by the growing plant.

455. **The Plant's Food** must contain these three elements in some shape or other. Let us look for them in the materials which the plant is constantly taking from the soil and the air.

456. *Water* is the substance of which it takes in vastly more than of anything else : we well know how necessary it is to vegetable life. The plant imbibes water by the roots, which are specially constructed for taking it in, as a liquid when the soil is wet, and probably also in the form of vapor when the soil is only damp. That water in the form of vapor is absorbed by the leaves likewise, when the plant needs it, is evident from the way partly wilted leaves revive and freshen when sprinkled or placed in a moist atmosphere. Now water is composed of *hydrogen* and *oxygen*, two of the three elements of cellulose or plant-fabric. Moreover, the hydrogen and the oxygen exist in water in exactly the same proportions that they do in cellulose : so it is clear that water furnishes these two elements.

457. We inquire, therefore, after the third element, *carbon*. This is the same as pure charcoal. Charcoal is the carbon of a vegetable left behind after charring, that is, heating it out of contact of the air until the hydrogen and oxygen are driven off. The charcoal of wood is so abundant in bulk as to preserve perfectly the shape of the cells after charring, and in weight it amounts to about half that of the original material. Carbon itself is a solid, and not at all dissolved by water: as such, therefore, it cannot be absorbed into the plant, however minute the particles ; only liquid and air can pass through the walls of the cells (402, 410). It must therefore come to the plant in some combination, and in a fluid form. The only substance within the plant's reach containing carbon in the proper state is

458. *Carbonic Acid.* This is a gas, and one of the components of the atmosphere, everywhere making about $\frac{1}{2500}$ part of its bulk, —enough for the food of plants, but not enough to be injurious to animals. For when mixed in any considerable proportion with the air we breathe, carbonic acid is very poisonous. The air produced by burning charcoal is carbonic acid, and we know how soon burning charcoal in a close room will destroy life.

459. The air around us consists, besides this minute proportion of carbonic acid, of two other gases, mixed together, viz. *oxygen*

and *nitrogen*. The nitrogen gas does not support animal life · it only dilutes the oxygen, which does. It is the oxygen gas alone which renders the air fit for breathing.

460. Carbonic acid consists of carbon combined with oxygen. In breathing, animals are constantly forming carbonic acid gas by uniting carbon from their bodies with oxygen of the air; they inspire oxygen into their lungs; they breath it out as carbonic acid. So with every breath animals are diminishing the oxygen of the air, — so necessary to animal life, — and are increasing its carbonic acid, — so hurtful to animal life ; or rather, which would be so hurtful if it were allowed to accumulate in the air. The reason why it does not increase in the air beyond this minute proportion is that plants feed upon it. They draw their whole stock of carbon from the carbonic acid of the air.

461. Plants take it in by their leaves. Every current, or breeze that stirs the foliage, brings to every leaf a succession of fresh atoms of carbonic acid, which it absorbs through its thousands of breathing-pores. We may prove this very easily, by putting a small plant or a fresh leafy bough into a glass globe, exposed to sunshine, and having two openings, causing air mixed with a known proportion of carbonic acid gas to enter by one opening, slowly traverse the foliage, and pass out by the other into a vessel proper to receive it : now, examining the air chemically, it will be found to have less carbonic acid than before. A portion has been taken up by the foliage.

462. Plants also take it in by their roots, some probably as a gas, in the same way that leaves absorb it, and much, certainly, dissolved in the water which the rootlets imbibe. The air in the soil, especially in a rich soil, contains many times as much carbonic acid as an equal bulk of the atmosphere above. Decomposing vegetable matter or manures, in the soil, are constantly evolving carbonic acid, and a large part of it remains there, in the pores and crevices, among which the absorbing rootlets spread and ramify. Besides, as this gas is dissolved by water in a moderate degree, every rain-drop that falls from the clouds to the ground brings with it a little carbonic acid, dissolving or washing it out of the air as it passes, and bringing it down to the roots of plants. And what flows off into the streams and ponds serves for the food of water-plants.

463. So water and carbonic acid, taken in by the leaves, or taken in by the roots and carried up to the leaves as crude sap, are the general food of plants, — are the raw materials out of which at least

14 *

the fabric and a part of the general products of the plant are made. Water and carbonic acid are *mineral matters:* in the plant, mainly in the foliage, they are changed into *organic matters.* This is

464. **The Plant's proper Work, Assimilation,** viz. the conversion by the vegetable of foreign, dead, mineral matter into its own living substance, or into organic matter capable of becoming living substance. To do this is, as we have said, the peculiar office of the plant. How and where is it done?

465. *It is done in the green parts of plants alone, and only when these are acted upon by the light of the sun.* The sun in some way supplies a power which enables the living plant to originate these peculiar chemical combinations, — to organize matter into forms which are alone capable of being endowed with life. The proof of this proposition is simple ; and it shows at the same time, in the simplest way, what the plant does with the water and carbonic acid it consumes. Namely, 1st, it is only in sunshine or bright daylight that the green parts of plants give out oxygen gas, — then they do ; and 2d, the giving out of this oxygen gas is just what is required to render the chemical composition of water and carbonic acid the same as that of *cellulose* (454), that is, of the plant's fabric. This shows why plants spread out so large a surface of foliage.

466. In plants growing or placed under water we may see bubbles of air rising from the foliage ; we may collect enough of this air to test it by a candle's burning brighter in it ; which shows it to be oxygen gas. Now if the plant is making cellulose or plant-substance, — that is, is making the very materials of its fabric and growth, as must generally be the case, — all this oxygen gas given off by the leaves comes from the decomposition of carbonic acid taken in by the plant.

467. This *must be so,* because cellulose is composed of 5 parts of oxygen and 10 of hydrogen to 6 of carbon (454) : here the first two are just in the same proportions as in water, which consists of 1 part of oxygen and 2 of hydrogen,—so that 5 parts of water and 6 of carbon represent 1 of cellulose or plant-fabric ; and to make it out of water and carbonic acid, the latter (which is composed of carbon and oxygen) has only to give up all its oxygen. In other words, the plant, in its foliage under sunshine, decomposes carbonic acid gas, and turns the carbon together with water into cellulose, at the same time giving off the oxygen of the carbonic acid into the air.

468. And we can readily prove that it is so,—namely, that plants

do decompose carbonic acid in their leaves and give out its oxygen, — by the experiment mentioned in paragraph 461. There the leaves, as we have stated, are taking in carbonic acid gas. We now add, that they are giving out oxygen gas at the same rate. The air as it comes from the glass globe is found to have just as much more oxygen as it has less carbonic acid than before — just as much more oxygen as would be required to turn the carbon retained in the plant back into carbonic acid again.

469. It is all the same when plants — instead of making fabric at once, that is, growing — make the prepared material, and store it up for future use. The principal product of plants for this purpose is *Starch*, which consists of minute grains of organic matter, lying loose in the cells. Plants often accumulate this, perhaps in the root, as in the Turnip, Carrot, and Dahlia (Fig. 57 – 60) ; or in subterranean stems or branches, as in the Potato (Fig. 68), and many rootstocks ; or in the bases of leaves, as in the Onion, Lily (Fig. 73 – 75), and other bulbs ; or in fleshy leaves above ground, as those of the Ice-Plant, House-leek, and Century-Plant (Fig. 82) ; or in the whole thickened body, as in many Cactuses (Fig. 76) ; or in the seed around the embryo, as in Indian Corn (Fig. 38, 39) and other grain ; or even in the embryo itself, as in the Horsechestnut (Fig. 23, 24), Bean (Fig. 16), Pea (Fig. 19), &c. In all these forms this is a provision for future growth, either of the plant itself or of some offset from it, or of its offspring, as it springs from the seed. Now starch is to cellulose or vegetable fabric just what the prepared clay is to the potter's vessel, — the same thing, only requiring to be shaped and consolidated. It has exactly the same chemical composition, and is equally made of carbon and the elements of water, by decomposing the same amount of carbonic acid and giving back its oxygen to the air. In using it for growth, the plant dissolves it, conveys it to the growing parts, and consolidates it into fabric.

470. *Sugar*, another principal vegetable product, also has essentially the same chemical composition, and may be formed out of the same common food of plants, with the same result. The different kinds of sugar (that of the cane, &c. and of grapes) consist of the same three materials as starch and cellulose, only with a little more water. The plant generally forms the sugar out of starch, changing one into the other with great ease ; starch being the form in which prepared material is stored up, and sugar that in which it is ex-

pended or transferred from one part of the plant to another. In the Sugar-cane and Indian Corn, starch is deposited in the seed ; in germination this is turned into sugar for the plantlet to begin its growth with ; the growing plant produces more, and deposits some as starch in the stalk ; just before blossoming, this is changed into sugar again, and dissolved in the sap, to form and feed the flowers (which cannot, like the leaves, create nourishment for themselves) ; and what is left is deposited in the seed as starch again, with which to begin the same operation in the next generation.

471. We might enumerate other vegetable products of this class (such as oil, acids, jelly, the pulp of fruits, &c.), and show how they are formed out of the carbonic acid and water which the plant takes in. But those already mentioned are sufficient. In producing any of them, carbonic acid taken from the air is decomposed, its carbon retained, and its oxygen given back to the air. That is to say,

472. **Plants purify the Air for Animals,** by taking away the carbonic acid injurious to them, continually poured into it by their breathing, as well as by the burning of fuel and by decay, and restoring in its place an equal bulk of life-sustaining oxygen (460). And by the same operation, combining this carbon with the elements of water, &c., and elaborating them into organic matter, — especially into starch, sugar, oil, and the like, —

473. **Plants produce all the Food and Fabric of Animals.** The herbivorous animals feed directly upon vegetables ; and the carnivorous feed upon the herbivorous. Neither the one nor the other originate any organic matter. They take it all ready-made from plants, — altering the form and qualities more or less, and at length destroying or decomposing it.

474. Starch, sugar, and oil, for example, form a large part of the food of herbivorous animals and of man. When digested, they enter into the blood ; any surplus may be stored up for a time in the form of fat, being changed a little in its nature ; while the rest (and finally the whole) is decomposed into carbonic acid and water, and exhaled from the lungs in respiration ; — in other words, is given back to the air by the animal as the very same materials which the plant takes from the air as its food (463) ; — is given back to the air in the same form that it would have been if the vegetable matter had been left to decay where it grew, or if it had been set on fire and burned ; — and with the same result too as to the heat, the heat in this case producing and maintaining the proper temperature of the animal.

475. But starch, sugar, and the like, do not make any part of the flesh or fabric of animals. And that for the obvious reason, that they consist of only the three elements *carbon, hydrogen,* and *oxygen ;* whereas the flesh of animals has nitrogen as well as these three elements in its composition. The materials of the animal body, called *Fibrine* in the flesh or muscles, *Gelatine* in the sinews and bones, *Caseine* in the curd of milk, &c., are all forms of one and the same substance, composed of *carbon, hydrogen, oxygen,* and *nitrogen.* As nitrogen is a large constituent of the atmosphere, and animals are taking it into their lungs with every breath they draw, we might suppose that they take this element of their frame directly from the air. But they do not. Even this is furnished by vegetables, and animals receive it ready-made in their food. And this brings us to consider still another and most important vegetable product, of a different class from the rest (omitted till now, for the sake of greater simplicity) ; namely, what is called

476. *Proteine.* This name has been given to it by chemists, because it occurs under such a protean variety of forms. The *Gluten* of wheat and the *Legumine* of beans and other leguminous plants may be taken to represent it. It occurs in all plants, at least in young and growing parts. It does not make any portion of their tissue, but is contained in all living cells, as a thin jelly, mingled with the sap or juice, or as a delicate mucilaginous lining. In fact, it is formed earlier than the cell-wall itself, and the latter is moulded on it, as it were ; so it is also called *Protoplasm.* It disappears from common cells as they grow old, being transferred onward to new or forming parts, where it plays a very active part in growth. Mixed with starch, &c., it is accumulated in considerable quantity in wheat, beans, and other grains and seeds, especially those which are most nutritious as food. It is the proteine which makes them so nutritious. Taken by animals as food, it forms their flesh and sinews, and the animal part of their bones, without much change ; for it has the same composition,— is just the same thing, indeed, in some slightly different forms. To produce it, the plant employs, in addition to the carbonic acid and water already mentioned as its general food, some *ammonia ;* which is a compound of *hydrogen* and *nitrogen.* Ammonia (which is the same thing as hartshorn) is constantly escaping into the air in small quantities from all decomposing vegetable and animal substances. Besides, it is produced in every thunderstorm. Every flash of lightning causes some to be made (in the

form of *nitrate of ammonia*) out of the nitrogen of the air and the vapor of water. The reason why it never accumulates in the air so as to be perceptible is, that it is extremely soluble in water, as are all its compounds. So it is washed out of the atmosphere by the rain as fast as it is made or rises into it, and is brought down to the roots of plants, which take it in freely. When assimilated in the leaves along with carbon and water, proteine is formed, the very substance of the flesh of animals. So all flesh is vegetable matter in its origin.

477. Even the earthy matter of the bones, and the iron and other mineral matters in the blood of animals, are derived from the plants they feed upon, with hardly an exception. These are furnished by the earthy or mineral constituents of plants (452), and are merely accumulated in the animal frame.

478. Animals, therefore, depend absolutely upon vegetables for their being. The great object for which the All-wise Creator established the vegetable kingdom evidently is, that plants might stand on the surface of the earth between the mineral and the animal creations, and organize portions of the former for the sustenance of the latter.

LESSON XXVII.

PLANT-LIFE.

479. LIFE is known to us only by its effects. We cannot tell what it *is ;* but we notice some things which it *does.* One peculiarity of living things, which has been illustrated in the last Lesson, is their power of transforming matter into new forms, and thereby making products never produced in any other way. Life is also manifested by

480. **Motion,** that is, by self-caused movements. Living things move ; those not living are moved. Animals, living as they do upon organized food, — which is not found everywhere, — must needs have the power of going after it, of collecting it, or at least of taking it in ; which requires them to make spontaneous movements. But plants, with their wide-spread surface (34, 131) always in con-

tact with the earth and air on which they feed, — the latter and the most important of these everywhere just the same, — have no need of locomotion, and so are generally fixed fast to the spot where they grow.

481. Yet many plants move their parts freely, sometimes when there is no occasion for it that we can understand, and sometimes accomplishing by it some useful end. The sudden closing of the leaflets of the Sensitive Plant, and the dropping of its leafstalk, when jarred, also the sudden starting forwards of the stamens of the Barberry at the touch, are familiar examples. Such cases seem at first view so strange, and so different from what we expect of a plant, that these plants are generally imagined to be endowed with a peculiar faculty, denied to common vegetables. But a closer examination will show that plants generally share in this faculty; that similar movements may be detected in them all, only — like those of the hands of a clock, or of the shadow of a sun-dial — they are too slow for the motion to be directly seen.

482. It is perfectly evident, also, that growth requires motion; that there is always an internal activity in living plants as well as in animals, — a power exerted which causes their fluids to move or circulate, and carries materials from one part to another. Some movements are mechanical; but even these are generally directed or controlled by the plant. Others must be as truly self-caused as those of animals are. Let us glance at some of the principal sorts, and see what light they throw upon vegetable life.

483. **Circulation in Cells.** From what we know of the anatomy of plants, it is clear that they have no general circulation (like that of all animals except the lowest), through a system of vessels opening into each other (402, 410). But in plants each living cell carries on a circulation of its own, at least when young and active. This may be beautifully seen in the transparent stems of Chara and many other water-plants, and in the leaves of the Fresh-water Tape-Grass (Vallisneria), under a good microscope. Here the sap circulates, often quite briskly in appearance, (but the motion is magnified as well as the objects,) in a steady stream, just beneath the wall, around each cell, passing up one side, across the end, down the other, and so round to complete the circuit, carrying with it small particles, or the larger green grains, which make the current more visible. This circulation may also be observed in hairs, particularly those on flowers, such as the jointed hairs of Spiderwort, looking

under the glass like strings of blue beads, each bead being a cell. But here a microscope magnifying six or eight hundred times in diameter is needed to see the current distinctly.

484. The movement belongs to the *protoplasm* (476), or jelly-like matter under the cell-wall. As this substance has just the same composition as the flesh of animals, it is not so strange that it should exhibit such animal-like characters. In the simplest water-plants, of the Sea-weed family, the body which answers to the seed is at first only a rounded little mass of protoplasm. When these bodies escape from the mother plant, they often swim about freely in the water in various directions, by a truly spontaneous motion, when they closely resemble animalcules, and are often mistaken for them. After enjoying this active life for several hours, they come to rest, form a covering of cellulose, and therefore become true vegetable cells, fix themselves to some support, germinate, and grow into the perfect plant.

485. **Absorption, Conveyance of the Sap, &c.** Although contained in cells with closed walls, nevertheless the fluids taken in by the roots are carried up through the stem to the leaves even of the topmost bough of the tallest tree. And the sap, after its assimilation by the leaves, is carried down in the bark or the cambium-layer, and distributed throughout the plant, or else is conveyed to the points where growth is taking place, or is accumulated in roots, stems, or wherever a deposit is being stored up for future use (71, 104, 128, 469).

486. That the rise of the sap is pretty rapid in a leafy and growing plant, on a dry summer's day, is evident from the amount of water it is continually losing by exhalation from the foliage (447) ; — a loss which must all the while be supplied from the roots, or else the leaves would dry up and die ; as they do so promptly when separated from the stem, or when the stem is cut off from the roots. Of course they do not then lose moisture any faster than they did before the separation ; only the supply is no longer kept up from below.

487. The rise of the sap into the leaves apparently is to a great degree the result of a mode of diffusion which has been called *Endosmose*. It acts in this way. Whenever two fluids of different density are separated by a membrane, whether of dead or of living substance, or are separated by any porous partition, a flow takes place through the partition, mainly towards the heavier fluid, until that is brought to the same density as the other. A familiar illus-

tration is seen when we place powdered sugar upon strawberries, and slightly moisten them : the dissolving sugar makes a solution stronger than the juice in the cells of the fruit; so this is gradually drawn out.   Also when pulpy fruits are boiled in a strong sirup; as soon as the sirup becomes denser than the juice in the fruit, the latter begins to flow out and the fruit begins to shrivel.   But when shrivelled fruits are placed in weak sirup, or in water, they become plump, because the flow then sets inwards, the juice in the cells being denser than the water outside.   Now the cells of the living plant contain organic matter, in the form of mucilage, protoplasm, sometimes sugar, &c.; and this particularly abounds in young and growing parts, such as the tips of roots (Fig. 56), which, as is well known, are the principal agents in absorbing moisture from the ground.   The contents of their cells being therefore always much denser than the moisture outside (which is water containing a little carbonic acid, &c., and a very minute quantity of earthy matter), this moisture is constantly drawn into the root.   What makes it ascend to the leaves ?

488.   To answer this question, we must look to the leaves, and consider what is going on there.   For (however it may be in the spring before the leaves are out)̤ in a leafy plant or tree the sap is not forced up from below, but is drawn up from above.   Water largely evaporates from the leaves (447) ; it flies off into the air as vapor, leaving behind all the earthy and the organic matters, — these not being volatile ; — the sap in the cells of the leaf therefore becomes denser, and so draws upon the more watery contents of the cells of the stalk, these upon those of the stem below, and so on, from cell to cell down to the root, causing a flow from the roots to the leaves, which begins in the latter, — just as a wind begins in the direction towards which it blows.   Somewhat similarly, elaborated sap is drawn into buds or any growing parts, where it is consolidated into fabric, or is conveyed into tubers, roots, seeds, and the like, in which it is condensed into starch and stored up for future use (74, 103, &c.).

489.   So in absorbing moisture by the roots, and in conveying the sap or the juices from cell to cell and from one part to another, the plant appears to make use of a physical or inorganic force ; but it manages and directs this as the purposes of the vegetable economy demand.   Now, when the proper materials are brought to the growing parts, *growth* takes place ; and in growth the plant moves

15

the particles of matter, arranges them, and shapes the fabric in a manner which we cannot at all explain by any mechanical laws. The organs are not shaped by any external forces; they shape themselves, and take such forms and positions as the nature of each part, or the kind of plant, requires.

490. **Special Movements.** Besides growing, and quite independent of it, plants not only assume particular positions, but move or bend one part upon another to do so. Almost every species does this, as well as what are called sensitive plants. In springing from the seed, the radicle or stem of the embryo, if not in the proper position already, bends itself round so as to direct its root-end downwards, and the stem-end or plumule upwards. It does the same when covered so deeply by the soil that no light can affect it, or when growing in a perfectly dark cellar. But after reaching the light, the stem bends towards that, as every one knows; and bends towards the stronger light, when the two sides are unequally exposed to the sun. It is now known that the shoot is bent by the shortening of the cells on the more illuminated side; for if we split the bending shoot in two, that side curves over still more, while the opposite side inclines to fly back. But how the light causes the cells to shorten on that side, we can no more explain, than we can tell how the will, acting through the nerves, causes the contraction of the fibres of the muscles by which a man bends his arm. We are sure that the bending of the shoot has nothing to do with growth, because it takes place after a shoot is grown; and the delicate stem of a young seedling will bend a thousand times faster than it grows. Also because it is yellow light that most favors growth and the formation of vegetable fabric, while the blue and violet rays produce the bending. Leaves also move, even more freely than stems. They constantly present their upper face to the light; and when turned upside down, they twist on their stalks, or curve round to recover their original position. The free ends of twining stems, as of Hop, or Morning Glory, or Bean, which apparently hang over to one side from their weight, are in fact bent over, and, the direction of the bend constantly changing, the shoot is steadily sweeping round the circle, making a revolution every few hours, or even more rapidly in certain cases, until it reaches a neighboring support, when, by a continuation of the same movement, it twines around it. Most tendrils revolve in the same way, sometimes even more rapidly; while others only turn from the

light; this is especially the case with those that cling to walls or trunks by sucker-like disks, as Virginia Creeper, p. 38, fig. 62. When an active tendril comes into contact with a stem or any such extraneous body, it incurves at the point of contact, and so lays hold of the support: the same contraction or tendency to curve affecting the whole length of the tendril, it soon shortens into a coil, part coiling one way, part the other, thus drawing the shoot up to the supporting body; or, if the tendril be free, it winds up in a simple coil. This movement of tendrils is so prompt in the Star-Cucumber (Sicyos) in Echinocystis, and in two sorts of Passion-flower, that the end, after a gentle rubbing, coils up by a movement rapid enough to be readily seen. In plants that climb by their leaf-stalks, such as Maurandia and Tropæolum, the movements are similar, but much too slow to be seen.

491. The so-called *sleep of plants* is a change of position as night draws on, and in different ways, according to the species, — the Locust and Wood-Sorrel turning down their leaflets, the Honey Locust raising them upright, the Sensitive Plant turning them forwards one over another; and the next morning they resume their diurnal position. One fact, among others, showing that the changes are not *caused* by the light, but by some power in the plant itself, is this. The leaves of the Sensitive Plant close long before sunset; but they expand again before sunrise, under much less light than they had when they closed. In several plants the leaves take the nocturnal position when brushed or jarred, — in the common Sensitive Plant very suddenly, in other sorts less quickly, in the Honey Locust a little too slowly for us to see the motion. The way in which blossoms open and close, some when the light increases, some when it diminishes, illustrates the same thing. The stamens of the Barberry, when touched at the base on the inner side, — as by an insect seeking for honey, or by the point of a pin, — make a sudden jerk forward, and in the process commonly throw some pollen upon the stigma, which stands a little above their reach.

492. In many of these cases we plainly perceive that a useful end is subserved. But what shall we say of the Venus's Fly-trap of North Carolina, growing where it might be sure of all the food a plant can need, yet provided with an apparatus for catching insects, and actually capturing them expertly by a sudden motion, in the manner already described (126, Fig. 81)? Or of the leaflets of the

Desmodium gyrans of the East Indies, spontaneously falling and rising by turns in jerking motions nearly the whole day long? We can only say, that plants are alive, no less than animals, and that it is a characteristic of living things to move.

⁎ CRYPTOGAMOUS OR FLOWERLESS PLANTS.

493. IN all the foregoing Lessons, we have had what may be called plants of the higher classes alone in view. There are others, composing the lower grades of vegetation, to which some allusion ought to be made.

494. Of this sort are Ferns or Brakes, Mosses, Liverworts, Lichens, Sea-weeds, and Fungi or Mushrooms. They are all classed together under the name of *Flowerless Plants*, or *Cryptogamous Plants;* the former epithet referring to the fact that they do not bear real *blossoms* (with stamens and pistils) nor *seeds* (with an embryo ready-formed within). Instead of seeds they have *spores*, which are usually simple cells (392). The name *Cryptogamous* means, of hidden fructification, and intimates that they may have something answering to stamens and pistils, although not the same; and this is now known to be the case with most of them.

495. Flowerless plants are so very various, and so peculiar in each family, that a volume would be required to illustrate them. Curious and attractive as they are, they are too difficult to be studied botanically by the beginner, except the Ferns, Club-Mosses, and Horse-tails. For the study of these we refer the student at once to the *Manual of the Botany of the Northern United States*, and to the *Field, Forest, and Garden Botany*. The structure and physiology of these plants, as well as of the Mosses, Liverworts, Lichens, Sea-weeds, and Fungi, are explained in the *Structural Botany, or Botanical Text Book*, and in other similar works. When the student has become prepared for the study, nothing can be more interesting than these plants of the lowest orders.

LESSON XXVIII.

SPECIES AND KINDS.

496. UNTIL now, we have been considering plants as to their structure and their mode of life. We have, as it were, been reading the biography of an individual plant, following it from the tiny seedling up to the mature and fruit-bearing herb or tree, and learning how it grows and what it does. The botanist also considers *plants as to their relationships*.

497. Plants and animals, as is well known, have two great peculiarities: 1st, they form themselves; and 2d, they multiply themselves. They reproduce themselves in a continued succession of

498. **Individuals** (3). Mineral things occur as *masses*, which are divisible into smaller and still smaller ones without alteration of their properties (391). But organic things (vegetables and animals) exist as *individual beings*. Each owes its existence to a parent, and produces similar individuals in its turn. So each individual is a link of a chain; and to this chain the natural-historian applies the name of

499. **Species.** All the descendants from the same stock therefore compose one species. And it was from our observing that the several sorts of plants or animals steadily reproduce themselves, — or, in other words, keep up a succession of similar individuals, — that the idea of species originated. So we are led to conclude that the Creator established a definite number of species at the beginning, which have continued by propagation, each after its kind.

500. There are few species, however, in which man has actually observed the succession for many generations. It could seldom be proved that all the White Pine trees or White Oaks of any forest came from the same stock. But observation having familiarized us with the general fact, that individuals proceeding from the same stock are essentially alike, we infer from their close resemblance that these similar individuals belong to the same species. That is, we infer it when the individuals are as much like each other as those are which we know to have sprung from the same stock.

501. We do not infer it from every resemblance; for there is the resemblance of *kind*, — as between the White Oak and the Red Oak,

15 *

and between the latter and the Scarlet Oak : these, we take for granted, have not originated from one and the same stock, but from three separate stocks. Nor do we deny it on account of every difference ; for even the sheep of the same flock, and the plants raised from peas of the same pod, may show differences, and such differences occasionally get to be very striking. When they are pretty well marked, we call them

Varieties. The White Oak, for example, presents two or three varieties in the shape of the leaves, although they may be all alike upon each particular tree. The question often arises, practically, and it is often hard to answer, whether the difference in a particular case is that of a variety, or is specific. If the former, we may commonly prove it to be so by finding such intermediate degrees of difference in various individuals as to show that no clear line of distinction can be drawn between them ; or else by observing the variety to vary back again, if not in the same individual, yet in its offspring. Our sorts of Apples, Pears, Potatoes, and the like, show us that differences which are permanent in the individual, and continue unchanged through a long series of generations when propagated by division (as by offsets, cuttings, grafts, bulbs, tubers, &c.), are not likely to be reproduced by seed. Still they sometimes are so : and such varieties are called

Races. These are strongly marked varieties, capable of being propagated by seed. Our different sorts of Wheat, Indian Corn, Peas, Radishes, &c., are familiar examples : and the races of men offer an analogous instance.

502. It should be noted, that all varieties have a *tendency* to be reproduced by seed, just as all the peculiarities of the parent tend to be reproduced in the offspring. And by selecting those plants which have developed or inherited any desirable peculiarity, keeping them from mingling with their less promising brethren, and selecting again the most promising plants raised from their seeds, we may in a few generations render almost any variety transmissible by seed, so long as we take good care of it. In fact, this is the way the cultivated or domesticated races, so useful to man, have been fixed and preserved. Races, in fact, can hardly, if at all, be said to exist independently of man. But man does not really produce them. Such peculiarities — often surprising enough — now and then originate, we know not how (the plant *sports*, as the gardeners say) ; they are only preserved, propagated, and generally further developed, by the culti-

vator's skilful care. If left alone, they are likely to dwindle and
perish, or else revert to the original form of the species.

503. Botanists variously estimate the number of known species
of plants at from seventy to one hundred thousand. About 3,850
species of the higher classes grow wild in the United States east of
the Mississippi. So that the vegetable kingdom exhibits a very
great diversity. Between our largest and highest-organized trees,
such as a Magnolia or an Oak, and the simplest of plants, reduced
to a single cell or sphere, much too minute to be visible to the
naked eye, how wide the difference! Yet the extremes are connected by intermediate grades of every sort, so as to leave no wide
gap at any place; and not only so, but every grade, from the most
complex to the most simple, is exhibited under a wide and most
beautiful diversity of forms, all based upon the one plan of vegetation which we have been studying, and so connected and so answering to each other throughout as to convince the thoughtful
botanist that all are parts of one system, works of one hand, realiza-
tions in nature of the conception of One Mind. We perceive this,
also, by the way in which the species are grouped into

504. Kinds. If the species, when arranged according to their re-
semblances, were found to differ from one another about equally, —
that is, if No. 1 differed from No. 2 just as much as No. 2 did from
No. 3, and No. 4 from No. 5, and so on throughout, — then, with all
the diversity in the vegetable kingdom there is now, there would yet
be no foundation in nature for grouping species into kinds. Species
and kinds would mean just the same thing. We should classify them,
no doubt, for convenience, but our classification would be arbitrary.
The fact is, however, that species resemble each other in very unequal degrees. Some species are almost exactly alike in their whole
structure, and differ only in the shape or proportion of their parts;
these, we say, belong to one *Genus*. Some, again, show a more general resemblance, and are found to have their flowers and seeds constructed on the same particular plan, but with important differences
in the details; these belong to the same *Order* or *Family*. Then,
taking a wider survey, we perceive that they all group themselves
under a few general types (or patterns), distinguishable at once by
their flowers, by their seeds or embryos, by the character of the
seedling plant, by the structure of their stems and leaves, and by
their general appearance: these great groups we call *Classes*.
Finally, we distinguish the whole into two great types or grades;

the higher grade of Flowering plants, exhibiting the full plan of vegetation, and the lower grade of Flowerless plants, in which vegetation is so simplified that at length the only likeness between them and our common trees or Flowering plants is that they are both vegetables. From species, then, we rise first to

505. **Genera** (plural of *Genus*). The Rose kind or genus, the Oak genus, the Chestnut genus, &c., are familiar illustrations. Each genus is a group of nearly related species, exhibiting a particular plan. All the Oaks belong to one genus, the Chestnuts to another, the Beech to a third. The Apple, Pear, and Crab are species of one genus, the Quince represents another, the various species of Hawthorn a third. In the animal kingdom the common cat, the wild cat, the panther, the tiger, the leopard, and the lion are species of the cat kind or genus; while the dog, the jackal, the different species of wolf, and the foxes, compose another genus. Some genera are represented by a vast number of species, others by few, very many by only one known species. For the genus may be as perfectly represented in one species as in several, although, if this were the case throughout, genera and species would of course be identical (504). The Beech genus and the Chestnut genus would be just as distinct from the Oak genus even if but one Beech and one Chestnut were known; as indeed was the case formerly.

506. **Orders or Families** (the two names are used for the same thing in botany) are groups of genera that resemble each other; that is, they are to genera what genera are to species. As familiar illustrations, the Oak, Chestnut, and Beech genera, along with the Hazel genus and the Hornbeams, all belong to one order, viz. the Oak Family; the Birches and the Alders make another family; the Poplars and Willows, another; the Walnuts (with the Butternut) and the Hickories, another. The Apple genus, the Quince and the Hawthorns, along with the Plums and Cherries and the Peach, the Raspberry, with the Blackberry, the Strawberry, the Rose, and many other genera, belong to a large order, the Rose Family.

507. **Tribes and Suborders.** This leads us to remark, that even the genera of the same order may show very unequal degrees of resemblance. Some may be very closely related to one another, and at the same time differ strikingly from the rest in certain important particulars. In the Rose Family, for example, there is the Rose genus itself, with the Raspberry genus, the Strawberry, the Cinquefoil, &c. near it, but by no means so much like it as they are like each

other : this group, therefore, answers to what is called a *Tribe ;* and the Rose itself stands for another tribe. But we further observe that the Apple genus, the Hawthorns, the Quince, and the June-berry, though of the same order, and nearly related among themselves, differ yet more widely from the Rose and its nearest relations; and so, on the other hand, do the Plum and Cherry, the Peach and the Almond. So this great Rose Family, or Order, is composed of three groups, of a more marked character than tribes, — groups which might naturally be taken for orders ; and we call them *Suborders.* But students will understand these matters best after a few lessons in studying plants in a work describing the kinds.

508. **Classes.** These are great assemblages of orders, as already explained (515). The orders of Flowering Plants are numerous, no less than 134 being represented in the Botany of the Northern United States ; but they all group themselves under two great classes. One class comprises all that have seeds with a monocotyledonous embryo (32), endogenous stems (423), and generally parallel-veined leaves (139) ; the other, those with dicotyledonous embryo, exogenous stems, and netted-veined leaves ; and the whole aspect of the two is so different that they are known at a glance.

509. Finally, these two classes together compose the upper *Series* or grade of *Flowering or Phænogamous Plants,* which have their counterpart in the lower *Series* of *Flowerless or Cryptogamous Plants,* — composed of three classes, and about a dozen orders.

510. The universal members of classification are CLASS, ORDER, GENUS, SPECIES, always standing in this order. When there are more, they take their places as in the following schedule, which comprises all that are generally used in a natural classification, proceeding from the highest to the lowest, viz. : —

Series,
  CLASS,
    Subclass,
      ORDER, or FAMILY,
        Suborder,
          Tribe,
            Subtribe,
              GENUS,
                Subgenus or Section,
                  SPECIES,
                    Variety.

LESSON XXIX.

BOTANICAL NAMES AND CHARACTERS.

511. PLANTS are *classified,* — i. e. are marshalled under their re-
spective classes, orders, tribes, genera, and species, — and they are
characterized, — that is, their principal characteristics or distinguishing marks are described or enumerated, in order that,

First, their resemblances or differences, of various degrees, may
be clearly exhibited, and all the species and kinds ranked next to
those they are most related to ; — and

Secondly, that students may readily ascertain the botanical names
of the plants they meet with, and learn their peculiarities, properties,
and place in the system.

512. It is in the latter that the young student is chiefly interested.
And by his studies in this regard he is gradually led up to a higher
point of view, from which he may take an intelligent survey of the
whole general system of plants. But the best way for the student
to learn the classification of plants (or Botany as a system), is to use
it, in finding out by it the name and the peculiarities of all the wild
plants he meets with.

513. **Names.** The botanical name of a plant, that by which a
botanist designates it, is the name of its genus followed by that of
the species. The name of the genus or kind is like the family name
or surname of a person, as *Smith,* or *Jones.* That of the species
answers to the baptismal name, as *John,* or *James.* Accordingly,
the White Oak is called botanically *Quercus alba ;* the first word, or
Quercus, being the name of the Oak genus ; the second, *alba,* that
of this particular species. And the Red Oak is named *Quercus
rubra ;* the Black-Jack Oak, *Quercus nigra ;* and so on. The botanical names are all in Latin (or are Latinized), this being the
common language of science everywhere ; and according to the
usage of that language, and of most others, the name of the species
comes after that of the genus, while in English it comes before it.

514. **Generic Names.** A plant, then, is named by two words. The
generic name, or that of the genus, is one word, and a substantive.
Commonly it is the old classical name, when the genus was known
to the Greeks and Romans ; as *Quercus* for the Oak, *Fagus* for the

Beech, *Córylus*, the Hazel, and the like. But as more genera became known, botanists had new names to make or borrow. Many are named from some appearance or property of the flowers, leaves, or other parts of the plant. To take a few examples from the early pages of the *Manual of the Botany of the Northern United States*, — in which the derivation of the generic names is explained. The genus *Hepatica*, p. 6, comes from the shape of the leaf resembling that of the liver. *Myosurus*, p. 10, means mouse-tail. *Delphinium*, p. 12, is from delphin, a dolphin, and alludes to the shape of the flower, which was thought to resemble the classical figures of the dolphin. *Zanthorhiza*, p. 13, is from two Greek words meaning yellow-root, the common name of the plant. *Cimicifuga*, p. 14, is formed of two Latin words, meaning, to drive away bugs, the same as its common name of Bugbane, the Siberian species being used to keep away such vermin. *Sanguinaria*, p. 26, is named from the blood-like color of its juice.

515. Other genera are dedicated to distinguished botanists or promoters of natural science, and bear their names: such are *Magnolia*, p. 15, which commemorates the early French botanist, Magnol, and *Jeffersonia*, p. 20, named after President Jefferson, who sent the first exploring expedition over the Rocky Mountains. Others bear the name of the discoverer of the plant in question; as, *Sarracenia*, p. 23, dedicated to Dr. Sarrazin of Quebec, who was one of the first to send our common Pitcher-plant to the botanists of Europe ; and *Claytonia*, p. 65, first made known by the early Virginian botanist Clayton.

516. **Specific Names.** The name of the species is also a single word, appended to that of the genus. It is commonly an adjective, and therefore agrees with the generic name in case, gender, &c. Sometimes it relates to the country the species inhabits; as, Claytonia *Virginica*, first made known from Virginia ; Sanguinaria *Canadensis*, from Canada, &c. More commonly it denotes some obvious or characteristic trait of the species; as, for example, in Sarracenia, our northern species is named *purpurea*, from the purple blossoms, while a more southern one is named *flava*, because its petals are yellow ; the species of Jeffersonia is called *diphylla*, meaning two-leaved, because its leaf is divided into two leaflets. Some species are named after the discoverer, or in compliment to a botanist who has made them known ; as, Magnolia *Fraseri*, named after the botanist Fraser, one of the first to find this species ; Ra-

worthia *Michauxii,* p. 65, named for the early botanist Michaux; and Polygala *Nuttallii,* in compliment to **Mr.** Nuttall, who described it under another name. Such names of persons are of course written with a capital initial letter. Occasionally some old substantive name is used for the species; as Magnolia *Umbrella,* p. 49, and Ranunculus *Flammula,* p. 41. These are also written with a capital initial, and need not accord with the generic name in gender, &c.

517. The name of a variety, when it is distinct enough to require any, is made on the same plan as that of the species, and is written after it; as, Ranunculus Flammula, variety *reptans,* p. 41 (i. e. the creeping variety), and R. abortivus, variety *micranthus,* p. 42, or the small-flowered variety of this species.

518. **Names of Groups.** The names of tribes, orders, and the like, are in the plural number, and are commonly formed by prolonging the name of a genus of the group taken as a representative of it. For example, the order of which the Buttercup or Crowfoot genus, *Ranunculus,* is the representative, takes from it the name of *Ranunculaceæ* (Manual, p. 34); meaning *Plantæ Ranunculaceæ* when written out in full, that is, Ranunculaceous Plants. This order comprises several tribes; one of which, to which Ranunculus itself belongs, takes the name of *Ranunculeæ;* another, to which the genus *Clematis,* or the Virgin's-Bower, belongs, takes accordingly the name of *Clematideæ;* and so on. So the term *Rosaceæ* (meaning Rosaceous plants) is the name of the order of which the Rose (*Rosa*) is the well-known representative; and *Roseæ* is the name of the particular tribe of it which comprises the Rose.

519. A few orders are named on a somewhat different plan. The great order *Leguminosæ,* for instance (Manual, p. 123), is not named after any genus in it; but the fruit, which is a legume (356), gives the name of *Leguminous Plants.* So, likewise, the order *Umbelliferæ* (Manual, p. 187) means Umbelliferous or Umbel-bearing Plants; and the vast order *Compositæ* (Manual, p. 215) is so named because it consists of plants whose blossoms are crowded into heads of the sort which were called "compound flowers" by the old botanists (277).

520. **Characters.** The brief description, or enumeration in scientific terms, of the principal distinctive marks of a species, genus, order, or other group, as given in botanical works, is called its *Character.* Thus, in the Manual, already referred to, at the begin-

ning, the character of the first great series is given; then that of
the first class, of the first subclass, and of the first division under it.
Then, after the name of the order, follows its character (the *ordinal*
character) : under the name of each genus (as, 1. *Clematis*, p. 35)
is added the *generic* character, or description of what essentially
distinguishes it ; and finally, following the name of each species, is
the *specific* character, a succinct enumeration of the points in which
it mainly differs from other species of the same genus. See, for
illustration, *Clematis Viorna*, p. 36, where the sentence immediately
following the name is intended to characterize that species from all
others like it.

521. Under this genus, and generally where we have several species of a genus, the species are arranged under *sections*, and these
often under *subsections*, for the student's convenience in analysis, —
the character or description of a section applying to all the species
under it, and therefore not having to be repeated under each species.
Under Clematis, also, are two sections with names, or sub-genera,
which indicates that they might almost be regarded as two distinct
genera. But these details are best understood by practice, in the
actual studying of plants to ascertain their name and place. And to
this the student is now ready to proceed.

LESSON XXX.

HOW TO STUDY PLANTS.

522. HAVING explained, in the two preceding Lessons, the general principles of Classification, and of Botanical Names, we may
now show, by a few examples, how the student is to proceed in
applying them, and how the name and the place in the system of an
unknown plant are to be ascertained.

523. We suppose the student to be provided with a hand *magni-*
fying-glass, and, if possible, with a *simple microscope*, i. e. with a
magnifying-glass, of two or more different powers, mounted on a
support, over a stage, holding a glass plate, on which small flowers
or their parts may be laid, while they are dissected under the microscope with the points of needles (mounted in handles), or divided

by a sharp knife. Such a microscope is not *necessary*, except for very small flowers; but it is a great convenience at all times, and is indispensable in studying the more difficult orders of plants.

524. We suppose the student now to have a work in which the plants of the country or district are scientifically arranged and described: if in the Southern Atlantic States, Dr. Chapman's *Flora of the Southern States;* if north of Carolina and Tennessee, Gray's *Manual of the Botany of the United States*, fifth edition; or, as covering the whole ground as to common plants, and including also all the common cultivated plants, Gray's *Field, Forest, and Garden Botany*, which is particularly arranged as the companion of the present work; that containing brief botanical descriptions of the plants, and this the explanation of their general structure, and of the technical terms employed in describing them. To express clearly the distinctions which botanists observe, and which furnish the best marks to know a plant by, requires a good many technical terms, or words used with a precise meaning. These, as they are met with, the student should look out in the Glossary at the end of this volume. The terms in common use are not so numerous as they would at first appear to be. With practice they will soon become so familiar as to give very little trouble. And the application of botanical descriptive language to the plants themselves, indicating all their varieties of form and structure, is an excellent discipline for the mind, equal, if not in some respects superior, to that of learning a classical language.

525. The following illustrations and explanations of the way to use the descriptive work are, first, for The *Field, Forest, and Garden Botany*, that being the one which will be generally used by beginners and classes. This and the *Lessons*, bound together in a single compact volume, will serve the whole purpose of all but advanced students, teachers, and working botanists. Thus equipped, we proceed to

526. **The Analysis of a Plant.** A Buttercup will serve as well as any. Some species or other may be found in blossom throughout nearly the whole spring and summer; and, except at the very beginning of the season, the fruit, more or less developed, may be gathered with the blossom. To a full knowledge of a plant the fruit is essential, although the name may almost always be ascertained without it. This common yellow flower being under examination, we are to refer the plant to its proper class and order or

family.   The families are so numerous, and so generally distinguishable only by a combination of a considerable number of marks that
the student must find his way to them by means of a contrivance
called an *Analytical Key.*   This Key begins on p. 12.

527. It takes note of the most comprehensive possible division of
plants, namely those "producing true flowers and seeds," and those
" not producing flowers, propagated by spores."   To the first of
these, the great series of PHÆNOGAMOUS or FLOWERING PLANTS,
the plant under examination obviously belongs.

528. This series divides into those " with wood in a circle, or in
concentric annual circles or layers around a central pith, netted-veined
leaves, and parts of the flower mostly in fives or fours," — to which
might be added the dicotyledonous embryo, but that in the present
case is beyond the young student's powers, even if the fruit were at
hand ; — and into those " with wood in separate threads scattered
through the diameter of the stem, not in a circle," also the " leaves
mostly parallel-veined, and parts of the flower almost always in
threes, never in fives."   Although the hollowness of the stem of the
present plant may obscure its internal structure, a practised hand,
by throwing the light through a thin cross section of the stem under
the glass, would make it evident that its woody bundles were all in
a circle near the circumference, yet this could hardly be expected
of an unassisted and inexperienced beginner.   But the two other
and very obvious marks, the netted-veined leaves, and the number
five in both calyx and corolla, certify at once that the plant belongs
to the first class, EXOGENOUS or DICOTYLEDONOUS PLANTS.

529. We should now look at the flower more particularly, so
as to make out its general
plan of structure, which we
shall need to know all about
as we go on.   We observe
that it has a calyx of 5
sepals, though these are apt
to fall soon after the blossom
opens ; that the 5 petals are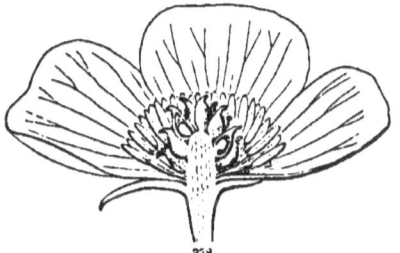
borne on the receptacle (or common axis of the flower) just above
the sepals and alternate with them ; that there are next borne, a

FIG. 358.   A flower of a Buttercup (Ranunculus bulbosus) cut through from top to bottom,
and enlarged.

little higher up on the receptacle, an indefinite number of stamens; and, lastly, covering the summit or centre of the receptacle, an in-

definite number of pistils. A good view of the whole is to be had by cutting the flower directly through the middle, from top to bottom (Fig. 358). If this be done with a sharp knife, some of the pistils will be neatly divided, or may be so by a second slicing. Each pistil, we see, is a closed ovary, containing a single ovule (Fig. 359) ascending from near the base of the cell, and is tipped with a very short broad style, which has the stigma running down the whole length of its inner edge. The ovary is little changed as it ripens into the sort of fruit termed an *akene* (Fig. 360); the ovule becoming the seed and fitting the cell (Fig. 361). Reverting to the key, on p. 13, we find that the class to which our plant belongs has two subclasses, one " with pistil of the ordinary sort, the ovules in a closed ovary"; the other "without proper pistil, the ovules naked on a scale," &c. The latter is nearly restricted to the Pine Family. The examination already had makes it quite clear that our plant belongs to the first subclass, ANGIOSPERMOUS Exogenous or Dicotyledonous Plants.

530. We have here no less than 110 orders under this subclass. To aid the unpractised student in finding his way among them, they are ranked under three artificial divisions; the *Polypetalous*, the *Monopetalous*, and the *Apetalous*. The plant in hand being furnished, in the words of the key, "with both calyx and corolla, the latter of wholly separate petals," is to be sought under I. POLY-PETALOUS DIVISION; for the analysis of which, see p. 14.

531. Fully half the families of the class rank under this division. The first step in the key is to the sections A and B; to the first of which, having " stamens more than 10, and more than twice the number of the sepals or divisions of the calyx," our plant must pertain.

532. Under this we proceed by a series of successive steps, their gradations marked by their position on the page, leading down to the name of the order or family, to which is appended the number

FIG. 359. A pistil taken from a Buttercup (Ranunculus bulbosus), and more magnified; its ovary cut through lengthwise, showing the ovule. 360. One of its pistils when ripened into a fruit (*achenium* or *akene*). 361. The same, cut through, to show the seed in it.

of the page where that family and the plants under it are described. The propositions of the same grade, two or more, from which determination is to be made, not only stand one directly under the other, but begin with the same word or phrase, or with some counterpart, — in the present case again with "Stamens," and with four propositions, with one and only one of which the flower in hand should agree. It agrees with the last of the four : "Stamens not monadelphous."

533. The propositions under this, to which we are now directed, are six, beginning with the word "Pistils" or "Pistil." The one which applies to the flower in hand is, clearly, the fourth : "Pistils numerous or more than one, separate, on the receptacle."

534. The terms of the analysis directly subordinate to this are only two : we have to choose between "Stamens borne on the calyx," and "Stamens borne on the receptacle." The latter is true of our flower. The terms subordinate to this are four, beginning with the word "Leaves." The fourth alone accords : "Leaves not peltate ; herbs," — and this line leads out to the CROWFOOT FAMILY, and refers to p. 33.

535. Turning to that page, a perusal of the brief account of the marks of the RANUNCULACEÆ (the technical Latin name) or CROW-FOOT FAMILY, assures us that the Key has led us safely and readily to a correct result. Knowing the order or family, we have next to ascertain the genus. Here are twenty genera to choose from ; but their characters are analyzed under sections and successive sub-sections (§, * , +, ++, &c.) so as to facilitate the way to the desired result. Of the two primary sections, we must reject § 1, as it agrees only in respect to the pistils, and differs wholly in the characters furnished by the sepals, the petals, and the leaves. With "§ 2. *Sepals imbricated in the bud: not climbing nor woody,*" it agrees. It also agrees with the sub-section immediately following, viz.: "* *Pistils and akenes, several or many in a head, one-seeded.*" The sub-division following : "+ *Petals none: sepals petal-like,*" is inapplicable ; but its counterpart, "+ + *Petals and sepals both conspicuous, five or more: akenes, naked. short-pointed,*" suits, and restricts our choice to the three genera, Adonis, Myosurus, and Ranunculus. The determination is soon made, upon noting the naked sepals, the petals with the little scale on the upper face of the short claw, and the akenes in a head : so the genus is, 7. RANUNCULUS.

16 *

536. The arrangement of the species of Ranunculus is to be found, under the proper number, 7, on p. 37 and the following. The first section contains aquatic species; ours is terrestrial, and in all other particulars answers to § 2. The smooth ovary and akene, and the perennial root refer it to the sub ection following, marked by the single star. The shape of the leaves excludes it from the " + Spearwort Crowfoots," the large and showy petals from the " + + Small-flowered Crowfoots; while all the marks agree with + + + BUTTERCUPS or COMMON CROWFOOTS. There is still a sub.division, one set marked, " ++ *Natives of the country, low or spreading*," the other " ++ ++ *Introduced weeds from Europe, common in fields, &c.: stem erect: leaves much cut*," — which is the case. We have then only to choose between the two field Crowfoots, and we have supposed the pupil to have in hand the lower, early-flowered one, common at the east, which has a solid bulb or corm at the base of the stem, and displays its golden flowers in spring or earliest summer, and which accordingly answers to the description of RANUNCULUS BULBOSUS, the BULBOUS BUTTERCUP.

537. Later in the season it might have been *R. acris*, the *Tall Buttercup*, or much earlier *R. fascicularis*, or *R. repens*. Having ascertained the genus from any one species, the student would not fail to recognize it again in any other, at a glance.

538. If now, with the same plant in hand, the *Manual* (Fifth edition) be the book used, the process of analysis will be so similar, that a brief indication of the steps may suffice. Here the corresponding Analytical Key, commencing on p. 21, leads similarly to the first Series, Class, Subclass, and Division ; — to A, with numerous stamens; 1, with calyx entirely free and separate from the pistil or pistils, thence to the fourth line beginning with the word Pistils; thence to the third of the three subordinate propositions, viz. to " Stamens inserted on the receptacle "; to the second of the succeeding couplet, or " Filaments longer than the anther "; to the second of the next couplet, " Flowers perfect," &c., and to the first of the final couplet, " Leaves not peltate ; petals deciduous," — which ends in " RANUNCULACEÆ, 34." This is the technical name of the family, and the page where it is described.

539. Turning to that page we read the general description of that order, particularly the portion at the beginning printed in *italics*, which comprises the more important points. The " Synopsis of the

Genera " which follows is similar to, but more technical than that of the other, more elementary book ; and the names of the tribes or natural groups of genera (507) are inserted. The steps of analysis bring the student to the Tribe III. RANUNCULEÆ, and under it to the genus RANUNCULUS. The number prefixed to the name enables the student to turn forward and find the genus, p. 40. The name, scientific and popular, is here followed by a full generic character (520). The primary sections here have names: the plant under examination belongs to "§ 2. RANUNCULUS proper"; and thence is to be traced, through the subdivisions *, + + + +, ++ ++, to the ultimate subdivision *b.*, under which, through a comparison of characters, the student reaches the species R. BULBOSUS, L.

540. The L. at the end of the name is the recognized abbreviation of the name of Linnæus, the botanist who gave it. Then come the common or English names ; then the specific character ; after this, the station where the plant grows, and the region in which it occurs. This is followed by the time of blossoming (from May to July); and then by some general descriptive remarks. The expression " Nat. from Eu." means that the species is a naturalized emigrant from Europe, and is not original to this country. But all these details are duly explained in the Preface to the *Manual,* which the student who uses that work will need to study.

LESSON XXXI.

HOW TO STUDY PLANTS : FURTHER ILLUSTRATIONS.

541. BEGINNERS should not be discouraged by the slow progress they must needs make in the first trials. By perseverance the various difficulties will soon be overcome, and each successful analysis will facilitate the next. Not only will a second species of the same genus be known at a glance, but commonly a second genus of the same order will be recognized as a relative at sight, by the family likeness. Or if the family likeness is not detected at the first view, it will be seen as the characters of the plant are studied out.

542. To help on the student by a second example, we will take the common cultivated Flax. Turning to the Key, as before, on

p. 12, the student is led to ask, first, is the plant PHÆNOGAMOUS or
FLOWERING? Of course it is; the blossom, with its
stamens and pistils, answers that question. Next, to
which of the two classes of Flowering Plants does it
belong? If we judge by the stem, we ask whether it
is exogenous or endogenous (422–424). A section of
the stem, considerably magnified, given on page 151,
we may here repeat (Fig. 362); it plainly shows a
ring of wood between a central pith and a bark. It is therefore
exogenous. Moreover, the leaves are netted-veined, though the
veins are not conspicuous. We might even judge from the embryo;
for there is little difficulty in dissecting a flax-seed, and in finding
that almost the whole interior is occupied by an embryo with two
cotyledons, much like that of an apple-seed (Fig. 11, 12), and this
class, as one of its name denotes, is dicotyledonous. If we view the
parts of the blossom, we perceive they are five throughout (Fig. 363,
365), a number which occurs in that class only. All these marks,
or as many of them as the student is able to verify, show that the
plant belongs to Class I. EXOGENOUS or DICOTYLEDONOUS PLANTS.

543. To which subclass, is the next inquiry. The single but
several-celled ovary in the centre of the flower, enclosing the ovules,
assures us that it belongs to the ANGIOSPERMOUS subclass, p. 13.

544. To get a good idea of the general plan of the flower, before

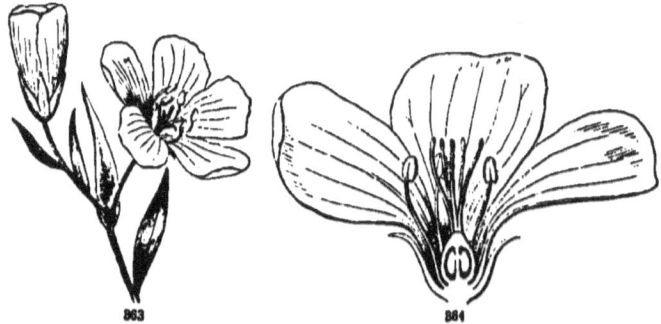

proceeding farther, cut it through the middle lengthwise, as in Fig
364, and also take a slice across a flower-bud, which will bring to view
an arrangement somewhat like that of Fig. 365. Evidently the
blossom is regularly constructed upon the number five. It has a
calyx of five sepals, a corolla of five petals, five stamens, and five

FIG. 362. Section of the stem of Flax, magnified. 363. Summit of a branch of the common
Flax, with two flowers. 364. A flower divided lengthwise and enlarged.

styles, with their ovaries all combined into one compound ovary. We note, also, that the several parts of the blossom are all free and unconnected, — the leaves of the calyx, the petals, and the stamens all rising separately one after another from the receptacle underneath the ovary; but the filaments, on close inspection, may show a slight union among themselves, at the base.

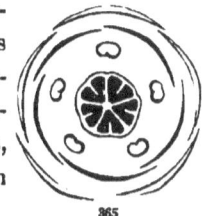

365

545. So our plant, having 5 separate petals, is of the POLYPETALOUS division of the first class, for the analysis of which see page 14.

546. But it does not belong to the primary division A, which has more than 10 stamens. The student passes on, therefore, to the counterpart division B, on page 16, to which the few stamens, here only five, refer it.

547. Of the three subdivisions, with numerals prefixed, only the second answers; for the calyx is free from the ovary, and there is only one ovary, although the styles are five.

548. The divisions subordinate to this form a couplet; and our plant agrees with the second member of it, having " Stamens of the same number as the petals " [5] and "alternate with them." The division under this is a triplet, of which we take the third member; for the " Leaves are not punctate with pellucid dots." Under this, in turn, is a triplet beginning with the word Ovary, and the five, if not ten cells, determine our choice of the third member of it, "Ovary compound." Under this we have no less than nine choices, dependent upon the structure of the ovary, the number of ovules and seeds, &c. But the 5-celled ovary with a pair of ovules in each cell, separated by a false partition projecting from the back (Fig. 365), so that the pod becomes in fact 10-celled, with a solitary seed in each cell, is described only in the ninth and last of the set, p. 18. Under this, again, we have to choose among five propositions relating to the seeds. Here the fifth — " Seeds and ovules only one or two in each cell " — alone meets the case. Under this, finally, we have to choose from six lines, beginning with the words Tree, Shrubs, or Herbs. The fifth alone agrees, and leads to the FLAX FAMILY, p. 77.

549. There is only one genus of it in this country, namely, the FLAX genus itself, or LINUM. To determine the species, look first

FIG. 365. Cross-section of an unexpanded flower of the same, a sort of diagram.

at the three sections, marked·with stars. The second answers to
our plant; and the annual root, pointed sepals, and blue petals determine it to be the COMMON FLAX, LINUM USITATISSIMUM.

550. By the *Manual*, the same plant would be similarly traced,
along a somewhat different order of steps, down to the genus on
p. 104, and to the species, which being a foreign cultivated one, and
only by chance spontaneous, is merely mentioned at the close.

551. After several analyses of this kind, the student will be able
to pass rapidly over most of these steps; should ordinarily recognize the class and the division at a glance. Suppose a common Mallow to be the next subject. Having flowers and seeds, it is Phænogamous. The netted-veined leaves, the structure of the stem, and
the leaves of the flower in fives, refer it to Class I. The pistils, of
the ordinary sort, refer it to Subclass I. The five petals refer it to
the Polypetalous division. Turning to the Key in the *Field, Forest,
and Garden Botany,* and to the analysis of that division, commencing
on p. 14, the numerous stamens fix it upon A, under which the
very first line, " Stamens monadelphous, united with the base of
the corolla; anthers kidney-shaped, one-celled," exactly expresses
the structure of these organs in our plant, which is thus determined
to be of the MALLOW FAMILY, — for which see page 70.

552. After reading the character of the family, and noting its
agreement in all respects, we fix upon § 1, in which the anthers are
all borne at the top, and not down the side of the tube of filaments.
We pass the subdivision with a single star, and choose the alternative,
with two stars, on account of the ring of ovaries, &c.; fix upon the
division ←, on account of the stigmas running down one side of the
slender style, instead of forming a little head or blunt tip at their
apex; and then have to choose among five genera. The three
separate bracts outside of the calyx, the obcordate petals, and the
fruit determine the plant to be a MALVA. Then, referring to p. 71
for the species, the small whitish flowers point to the first division,
and a comparison of the characters of the two species under it,
assures us that the plant in hand is MALVA ROTUNDIFOLIA.

553. For the sake of an example in the Monopetalous Division,
we take a sort of Morning-Glory which is often met with climbing
over shrubs along the moist banks of streams. Its netted-veined
leaves, the sepals and the stamens being five, — also the structure of
the stem, if we choose to examine it, and the embryo with two leafy

cotyledons (as in Fig. 26), readily inspected if we have seeds, — show it belongs to Class I. Its pistil refers it of course to Subclass I. The corolla being a short funnel-shaped tube, theoretically regarded as formed of five petals united up to the very summit or border, renders the flower a good illustration of the MONOPETALOUS DIVISION, the analysis of which begins on p. 20, in the work we are using.

554. The calyx free from the ovary excludes it from the section A, and refers it to section B. This is subdivided, in the first place, by the number of the stamens, and their position as respects the lobes of the corolla. Now, as the petals of the corolla in this flower are united up to the very border, the student may at first be puzzled to tell how many lobes it should have, or, in other words, how many petals enter into its composition. But the five leaves of the calyx would lead one to expect a corolla of five parts also. And, although there are here really no lobes or notches to be seen, yet the five plaits of the corolla answer to the notches, and show it to consist of five petals perfectly united. Since the stamens are of the same number as the plaits of the corolla, and are placed before them (as may be best seen by splitting down the corolla on one side and spreading it out flat), it follows that they alternate with the lobes or petals; therefore our plant falls under the third subdivision: "Stamens as many as the lobes or parts of the corolla and alternate with them." This subdivides by the pistils. Our plant, having a pistil with two stigmas and two cells to the ovary, must be referred to the fifth and last category: "Pistil one, with a single compound ovary," &c. We are then directed to the stamens, which here are "plainly borne on the corolla"; next to the leaves, which are on the stem (not all at the root), also alternate, without stipules; the stamens 5, and the ovary 2-celled, — all of which accords with the seventh of the succeeding propositions, and with no other. The middle one alone under this agrees as to the ovary and seeds, and all is confirmed by the twining stem. It is the CONVOLVULUS FAMILY. p. 262.

555. The proper Convolvulus Family has green foliage, as has our plant. Its style is single and entire, as in § 1. Its calyx has a pair of large leafy bracts, as in the subdivision with two stars. So we reach the genus CALYSTEGIA, or BRACTED BINDWEED.

556. Under this genus two species are described: the twining stem, and the other particulars of our plant, direct us to the first C. SEPIUM, which in England is named HEDGE BINDWEED, and here is one of the various Convolvulaceous plants known as MORNING-GLORY.

LESSON XXXII.

HOW TO STUDY PLANTS: FURTHER ILLUSTRATIONS.

557. The foregoing illustrations have all been of the first or Exogenous class. We will take one from the other class, and investigate it by the *Manual*.

558. It shall be a rather common plant of our woods in spring, the Three-leaved Nightshade, or Birthroot. With specimens in hand, and the *Manual* open at the Analytical Key, p. 21, seeing that the plant is of the Phænogamous series, we proceed to determine the class. The netted-veined leaves would seem to refer the plant to the first class; while the blossom (Fig. 366, 367), constructed on the number three, naturally directs us to the second

class, in which this number almost universally prevails. Here the student will be somewhat puzzled. If the seeds were ripe, they might be examined, to see whether the embryo has one cotyledon only, or a pair. But the seeds are not to be had in spring, and if they were, the embryo would not readily be made out. We must judge, therefore, by the structure of the stem. Is it exogenous or endogenous? If we cut the stem through, or take off a thin slice crosswise and lengthwise, we shall

perceive that the woody matter in it consists of a number of threads, interspersed throughout the soft cellular part without regularity, and not collected into a ring or layer. In fact, it is just like the Corn-stalk (Fig. 351), except that the woody threads are fewer. It is therefore *endogenous* (422); and this decides the question in favor of Class II. MONOCOTYLEDONOUS or ENDOGENOUS PLANTS (page 30), notwithstanding the branching veins of the leaves. For neither this character, nor the number of parts in

FIG. 366. Flower of Trillium erectum, viewed from above. 367. Diagram of the same, a cross-section of the unopened blossom, showing the number and arrangement of parts.

the blossom, holds good universally, while the plan of the stem does.

559. The single flower of our plant with distinct calyx and corolla takes us over the Spadiceous to the PETALOIDEOUS DIVISION: the Petaloideous Division of Endogens there begins on p. 28. These parts being free from and beneath the ovary, refer us to the third subdivision, viz : " 3. *Perianth wholly free from the ovary.*"

559*. The pistil is next to be considered : it accords with the third of the triplet : " Pistil one, compound (cells or placentæ 3) ; anthers 2-celled." Under this follows a triplet, of which the initial word is " Perianth ": our choice falls upon the first, as there is nothing " glumaceous " about this flower.

560. The succeeding triplet relates to the stamens ; here 6, so we take the first alternative. The next refers to mode and place of growth : our plant is " Terrestrial, and not rush-like." The next again to the perianth : the second number of the triplet : " Perianth of 3 foliaceous and green sepals, and 3 colored withering-persistent petals " (as would be seen after flowering-time), brings us to a particular group in the great Lily family, or LILIACEÆ, p. 520.

561. Reading over the family character, and collating the five tribes comprised, we perceive that our plant belongs to the group, quite peculiar among Liliaceous plants, here ranked as Tribe I. TRILLIDEÆ, the Trillium tribe. And the next step, leading to a choice between two genera, determines the genus to be TRILLIUM.

562. Turning to this, on p. 522, and reading the full description of it, we proceed to the easy task of ascertaining the species. The " flower is raised on a peduncle," as in § 2. This peduncle is slender and nearly erect, and all the other particulars accord with the subdivision marked by a single star. And, finally, the ovate, acutish, widely-spreading, dark dull-purple petals mark the species as the PURPLE BIRTHROOT, TRILLIUM ERECTUM, L.

563. By the *Field, Forest, and Garden Botany*, the analysis is similar, only more simple. The details need not be particularly recapitulated.

564. The student residing west of New England will also be likely to find another species, with similar foliage, but with larger, pure white, and obovate petals, turning rose-color when about to fade. This will at once be identified as *T. grandiflorum.* And towards the north, in cold and damp woods or swamps, a smaller

17

species will be met with, having dull-green and petioled leaves rounded at the base, and rather narrow, wavy, white petals, marked with pink or purple stripes at the base: this the student will refer to *T. erythrocarpum*. But the species principally found in the eastern parts of the country has a short peduncle recurved under the leaves, so as nearly to conceal the much less handsome, dull white flower: this, it will be seen, is *T. cernuum*, the *Nodding Trillium* or *Wake Robin*.

565. Whenever the student has fairly studied out one species of a genus, he will be likely to know the others when he sees them. And when plants of another genus of the same order are met with, the order may generally be recognized at a glance, from the family resemblance. For instance, having first become acquainted with the Convolvulus family in the genus Calystegia (555), we recognize it at once in the common Morning-Glory, and in the Cypress-Vine, and even in the Dodder, although these belong to as many different genera. Having examined the common Mallow (552), we immediately recognize the Mallow family (*Malvaceæ*) in the Marsh-Mallow, sparingly naturalized along the coast, in the Glade Mallow, and the Indian Mallow, in the Hibiscus or Rose-Mallow, and so of the rest: for the relationship is manifest in their general appearance, and in the whole structure of the flowers, if not of the foliage also.

566. So the study of one plant leads naturally and easily to the knowledge of the whole order or family of plants it belongs to: — which is a great advantage, and a vast saving of labor. For, although we have about one hundred and thirty orders of Flowering Plants represented in our Botany of the Northern States by about 2,540 species, yet half of these species belong to nine or ten of these orders; and more than four fifths of the species belong to forty of the orders. One or two hundred species, therefore, well examined, might give a good general idea of our whole botany. And students who will patiently and thoroughly study out twenty or thirty well-chosen examples will afterwards experience little difficulty in determining any of our Flowering Plants and Ferns, and will find the pleasure of the pursuit largely to increase with their increasing knowledge.

567. And the interest will be greatly enhanced as the student, rising to higher and wider views, begins to discern the *System* of Botany, or, in other words, comprehends more and more of *the Plan of the Creator in the Vegetable Kingdom*.

LESSON XXXIII.

BOTANICAL SYSTEMS.

568. **Natural System.** *The System* of Botany consists of the orders or families, duly arranged under their classes, and having the tribes, the genera, and the species arranged in them according to their relationships. This, when properly carried out, is the *Natural System;* because it is intended to express, as well as we are able, the various degrees of relationship among plants, as presented in nature; — to rank those species, those genera, &c. next to each other in the classification which are really most alike in all respects, or, in other words, which are constructed most nearly on the same particular plan.

569. Now this word *plan* of course supposes a *planner*, — an intelligent mind working according to a system: it is this system, therefore, which the botanist is endeavoring as far as he can to exhibit in a classification. In it we humbly attempt to learn something of the plan of the Creator in this department of Nature.

570. So there can be only *one* natural system of Botany, if by the term we mean the plan according to which the vegetable creation was called into being, with all its grades and diversities among the species, as well of past as of the present time. But there may be many natural systems, if we mean the attempts of men to interpret and express the plan of the vegetable creation, — systems which will vary with our advancing knowledge, and with the judgment and skill of different botanists, — and which must all be very imperfect. They will all bear the impress of individual minds, and be shaped by the current philosophy of the age: But the endeavor always is to make the classification a reflection of Nature, as far as any system can be which has to be expressed in a series of definite propositions, and have its divisions and subdivisions following each other in some single fixed order.*

---

* The best classification must fail to give more than an imperfect and considerably distorted reflection, not merely of the plan of creation, but even of our knowledge of it. It is often obliged to make arbitrary divisions where Nature shows only transitions, and to consider genera, &c. as equal units, or groups of equally related species, while in fact they may be very unequal, — to assume, on

571. The Natural System, as we receive it, and as to that portion of it which is represented in the botany of our country, is laid before the student in the *Manual of the Botany of the Northern United States*. The orders, however, still require to be grouped, according to their natural relationships, into a considerable number of great groups (or *alliances*) ; but this cannot yet be done throughout in any easy way. So we have merely arranged them somewhat after a customary order, and have given, in the *Artificial Key*, a contrivance for enabling the student easily to find the natural order of any plant. This is a sort of

572. **Artificial Classification.** The object of an artificial classification is merely to furnish a convenient method of finding out the name and place of a plant. It makes no attempt at arranging plants according to their relationships, but serves as a kind of dictionary. It distributes plants according to some one peculiarity or set of peculiarities (just as a dictionary distributes words according to their first letters), disregarding all other considerations.

573. At present we need an artificial classification in Botany only as a Key to the Natural Orders, — as an aid in referring an unknown plant to its proper family ; and for this it is very needful to the student. Formerly, when the orders themselves were not clearly made out, an artificial classification was required to lead the student down to the genus. Two such classifications were long in vogue. First, that of Tournefort, founded mainly on the leaves of the flower, the calyx and corolla : this was the prevalent system throughout the first half of the eighteenth century ; but it has long since gone by. It was succeeded by the well-known artificial system of Linnæus, which has been used until lately ; and which it is still worth while to give some account of.

574. **The Artificial System of Linnæus** was founded on the stamens and pistils. It consists of twenty-four classes, and of a variable number of orders, which were to take the place temporarily of the natural classes and orders ; the genera being the same under all classifications.

---

paper at least, a strictly definite limitation of genera, of tribes, and of orders, although observation shows so much blending here and there of natural groups, sufficiently distinct on the whole, as to warrant us in assuming the likelihood that the Creator's plan is one of *gradation, not of definite limitation,* even perhaps to the species themselves.

575. The twenty-four *classes* of Linnæus were founded upon something about the stamens. The following is an analysis of them. The first great division is into two great series, the *Phænogamous* and the *Cryptogamous*, the same as in the Natural System. The first of these is divided into those flowers which have the stamens in the same flower with the pistils, and those which have not ; and these again are subdivided, as is shown in the following tabular view.

Series I. PHÆNOGAMIA ; plants with stamens and pistils, i. e. with real flowers.

1. Stamens in the same flower as the pistils :
* Not united with them,
⊢ Nor with one another.
⊢⊢ Of equal length if either 6 or 4 in number.

One to each flower,	Class	1.	MONANDRIA.
Two " "		2.	DIANDRIA.
Three " "		3.	TRIANDRIA.
Four " "		4.	TETRANDRIA.
Five " "		5.	PENTANDRIA.
Six . " "		6.	HEXANDRIA.
Seven " "		7.	HEPTANDRIA.
Eight " "		8.	OCTANDRIA.
Nine " "		9.	ENNEANDRIA.
Ten " "		10.	DECANDRIA.
Eleven to nineteen to each flower,		11.	DODECANDRIA.
Twenty or more inserted on the calyx,		12.	ICOSANDRIA.
" " " on the receptacle,		13.	POLYANDRIA.

⊢⊢ Of unequal length and either 4 or 6.

Four, 2 long and 2 shorter,	14.	DIDYNAMIA.
Six, 4 long and 2 shorter,	15.	TETRADYNAMIA

⊢ ⊢ United with each other,
By their filaments,

Into one set or tube,	16.	MONADELPHIA.
Into two sets,	17.	DIADELPHIA.
Into three or more sets,	18.	POLYADELPHIA
By their anthers into a ring,	19.	SYNGENESIA.
* * United with the pistil,	20.	GYNANDRIA.

2. Stamens and pistils in separate flowers,

Of the same individuals,	21.	MONŒCIA.
Of different individuals,	22.	DIŒCIA.
Some flowers perfect, others staminate or pistillate either in the same or in different individuals,	23.	POLYGAMIA.

Series II. CRYPTOGAMIA. No stamens and pistils, therefore no proper flowers,   24. CRYPTOGAMIA

17 *

576. The names of these classes are all compounded of Greek words. The first eleven consist of the Greek numerals, in succession, from 1 to 11, combined with *andria*, which here denotes stamens; — e. g. *Monandria*, with one stamen; and so on. The 11th has the numeral for twelve stamens, although it includes all which have from eleven to nineteen stamens, numbers which rarely occur. The 12th means "with twenty stamens," but takes in any higher number, although only when the stamens are borne on the calyx. The 13th means "with many stamens," but it takes only those with the stamens borne on the receptacle. The 14th means "two stamens powerful," the shorter pair being supposed to be weaker; the 15th, "four powerful," for the same reason. The names of the next three classes are compounded of *adelphia*, brotherhood, and the Greek words for *one*, *two*, and *many* (*Monadelphia, Diadelphia,* and *Polyadelphia*). The 19th means "united in one household." The 20th is compounded of the words for stamens and pistils united. The 21st and 22d are composed of the word meaning *house* and the numerals *one*, or single, and *two* : *Monœcia*, in one house, *Diœcia*, in two houses. The 23d is fancifully formed of the words meaning *plurality* and *marriage*, from which the English word *polygamy* is derived. The 24th is from two words meaning *concealed nuptials*, and is opposed to all the rest, which are called *Phænogamous*, because their stamens and pistils, or parts of fructification, are evident.

577. Having established the classes of his system on the stamens, Linnæus proceeded to divide them into *orders* by marks taken from the pistils, for those of the first thirteen classes. These orders depend on the number of the pistils, or rather on the number of styles, or of stigmas when there are no styles, and they are named, like the classes, by Greek numerals, prefixed to *gynia*, which means *pistil*. Thus, flowers of these thirteen classes with

One style or sessile stigma belong to	Order 1.	MONOGYNIA.
Two styles or sessile stigmas, to	2.	DIGYNIA.
Three " "	3.	TRIGYNIA.
Four " "	4.	TETRAGYNIA.
Five " "	5.	PENTAGYNIA.
Six " "	6.	HEXAGYNIA.
Seven " "	7.	HEPTAGYNIA.
Eight " "	8.	OCTOGYNIA.
Nine " "	9.	ENNEAGYNIA.
Ten " "	10.	DECAGYNIA.
Eleven or twelve "	11.	DODECAGYNIA.
More than twelve "	13.	POLYGYNIA.

578. The orders of the remaining classes are founded on various considerations, some on the nature of the fruit, others on the number and position of the stamens. But there is no need to enumerate them here, nor farther to illustrate the Linnæan Artificial Classification. For as a system it has gone entirely out of use ; and as a Key to the Natural Orders it is not so convenient, nor by any means so certain, as a proper Artificial Key, prepared for the purpose, such as we have been using in the preceding Lessons.

·

LESSON XXXIV.

HOW TO COLLECT SPECIMENS AND MAKE AN HERBARIUM.

579. **For Collecting Specimens** the needful things are a large *knife*, strong enough to be used for digging up bulbs, small rootstocks, and the like, as well as for cutting woody branches; and a *botanical box*, or a *portfolio*, for holding specimens which are to be carried to any distance.

580. It is well to have both. The *botanical box* is most useful for holding specimens which are to be examined fresh. It is made of tin, in shape like a candle-box, only flatter, or the smaller sizes like an English sandwich-case ; the lid opening for nearly the whole length of one side of the box. Any portable tin box of convenient size, and capable of holding specimens a foot or fifteen inches long, will answer the purpose. The box should shut close, so that the specimens may not wilt : then it will keep leafy branches and most flowers perfectly fresh for a day or two, especially if slightly moistened.

581. The *portfolio* should be a pretty strong one, from a foot to twenty inches long, and from nine to eleven inches wide, and fastening with tape, or (which is better) by a leathern strap and buckle at the side. It should contain a quantity of sheets of thin and smooth, unsized paper ; the poorest printing-paper and grocers' tea-paper are very good for the purpose. The specimens as soon as gathered are to be separately laid in a folded sheet, and kept under moderate pressure in the closed portfolio.

582. Botanical specimens should be either in flower or in fruit. In the case of herbs, the same specimen will often exhibit the two; and both should by all means be secured whenever it is possible. Of small herbs, especially annuals, the whole plant, root and all, should be taken for a specimen. Of larger ones branches will suffice, with some of the leaves from near the root. Enough of the root or subterranean part of the plant should be collected to show whether the plant is an annual, biennial, or perennial. Thick roots, bulbs, tubers, or branches of specimens intended to be preserved, should be thinned with a knife, or cut into slices lengthwise.

583. **For drying Specimens** a good supply of soft and unsized paper — the more bibulous the better — is wanted ; and some convenient means of applying pressure. All that is requisite to make good dried botanical specimens is, to dry them as rapidly as possible between many thicknesses of paper to absorb their moisture, under as much pressure as can be given without crushing the more delicate parts. This pressure may be given by a botanical press, of which various forms have been contrived ; or by weights placed upon a board, — from forty to eighty or a hundred pounds, according to the quantity of specimens drying at the time. For use while travelling, a good portable press may be made of thick binders' boards for the sides, holding the drying paper, and the pressure may be applied by a cord, or, much better, by strong straps with buckles.

584. For drying paper, the softer and smoother sorts of cheap wrapping-paper answer very well. This paper may be made up into *driers*, each of a dozen sheets or less, according to the thickness, lightly stitched together. Specimens to be dried should be put into the press as soon as possible after gathering. If collected in a portfolio, the more delicate plants should not be disturbed, but the sheets that hold them should one by one be transferred from the portfolio to the press. Specimens brought home in the botanical box must be laid in a folded sheet of the same thin, smooth, and soft paper used in the portfolio ; and these sheets are to hold the plants until they are dry. They are to be at once laid in between the driers, and the whole put under pressure. Every day (or at first even twice a day would be well) the specimens, left undisturbed in their sheets, are to be shifted into well-dried fresh driers, and the pressure renewed, while the moist sheets are spread out to dry, that they may take their turn again at the next shifting. This course must be continued until the specimens are no longer moist to the touch, —

which for most plants requires about a week ; then they may be transferred to the sheets of paper in which they are to be preserved. If a great abundance of drying-paper is used, it is not necessary to change the sheets every day, after the first day or two.

585. **Herbarium.** The botanist's collection of dried specimens, ticketed with their names, place, and time of collection, and systematically arranged under their genera, orders, &c., forms a *Hortus Siccus* or *Herbarium.* It comprises not only the specimens which the proprietor has himself collected, but those which he acquires through friendly exchanges with distant botanists, or in other ways. The specimens of an herbarium may be kept in folded sheets of neat, and rather thick, white paper; or they may be fastened on half-sheets of such paper, either by slips of gummed paper, or by glue applied to the specimens themselves. Each sheet should be appropriated to one species ; two or more different plants should never be attached to the same sheet. The generic and specific name of the plant should be added to the lower right-hand corner, either written on the sheet, or on a ticket pasted down at that corner; and the time of collection, the locality, the color of the flowers, and any other information which the specimens themselves do not afford, should be duly recorded upon the sheet or the ticket. The sheets of the herbarium should all be of exactly the same dimensions. The herbarium of Linnæus is on paper of the common foolscap size, about eleven inches long and seven wide. But this is too small for an herbarium of any magnitude. Sixteen and a half inches by ten and a half, or eleven and a half inches, is an approved size.

586. The sheets containing the species of each genus are to be placed in *genus-covers,* made of a full sheet of thick, colored paper (such as the strongest Manilla-hemp paper), which fold to the same dimensions as the species-sheet ; and the name of the genus is to be written on one of the lower corners. These are to be arranged under the orders to which they belong, and the whole kept in closed cases or cabinets, either laid flat in compartments, like large "pigeon-holes," or else placed in thick portfolios, arranged like folio volumes, and having the names of the orders lettered on the back.

GLOSSARY

OR

DICTIONARY OF TERMS USED IN DESCRIB-
ING PLANTS,

COMBINED WITH AN INDEX.

A, at the beginning of words of Greek derivation, commonly signifies a negative, or the absence of something; as *apetalous*, without petals; *aphyllous*, leafless, &c. If the word begins with a vowel, the prefix is *an*; as *unantherous*, destitute of anther.

Abnormal: contrary to the usual or the natural structure.

Aboriginal: original in the strictest sense; same as indigenous.

Abortive: imperfectly formed, or rudimentary, as one of the stamens in fig. 195 and three of them in fig. 196, p. 95.

Abortion: the imperfect formation, or non-formation, of some part.

Abrupt: suddenly terminating; as, for instance,

Abruptly pinnate: pinnate without an odd leaflet at the end; fig. 128, p. 65.

Acaulescent (acaulis): apparently stemless; the proper stem, bearing the leaves and flowers, being very short or subterranean, as in Bloodroot, and most Violets; p. 36.

Accessory: something additional; as *Accessory buds*, p. 26.

Accrescent: growing larger after flowering, as the calyx of Physalis.

Accumbent: lying against a thing. The cotyledons are accumbent when they lie with their edges against the radicle.

Acerose: needle-shaped, as the leaves of Pines; fig. 140, p. 72.

Acetabuliform: saucer-shaped.

Achenium (plural *achenia*) : a one-seeded, seed-like fruit; fig. 286, p. 129

Achlamydeous (flower) : without floral envelopes; as Lizard's-tail, p. 90, fig. 180.

Acicular: needle-shaped ; more slender than acerose.

Acinaciform: scymitar-shaped, like some bean-pods.

Acines: the separate grains of a fruit, such as the raspberry; fig. 289.

Acorn: the nut of the Oak ; fig. 299, p. 130.

Acotyledonous: destitute of cotyledons or seed-leaves.

Acrogenous: growing from the apex, as the stems of Ferns and Mosses.

Acrogens, or *Acrogenous Plants:* the higher Cryptogamous plants, such as Ferns, &c., p. 172.

Aculeate: armed with prickles, l. c. *aculei;* as the Rose and Brier.

Aculeolate: armed with small prickles, or slightly prickly.

Acuminate: taper-pointed, as the leaf in fig. 97 and fig. 103.

Acute: merely sharp-pointed, or ending in a point less than a right angle.

Adelphous (stamens): joined in a fraternity (*adelphia*): see *monadelphous* and *diadelphous.*

Adherent: sticking to, or, more commonly, growing fast to another body; p. 104.

Adnate: growing fast to; it means born adherent. The anther is adnate when fixed by its whole length to the filament or its prolongation, as in Tulip-tree, fig. 233.

Adpressed, or *appressed:* brought into contact, but not united.

Adscendent, ascendent, or *ascending:* rising gradually upwards.

Adsurgent, or *assurgent:* same as ascending.

Adventitious: out of the proper or usual place; e. g. *Adventitious buds,* p. 26, 27.

Adventive: applied to foreign plants accidentally or sparingly spontaneous in a country, but hardly to be called naturalized.

Æquilateral: equal-sided; opposed to oblique.

Æstivation: the arrangement of parts in a flower-bud, p. 108.

Air-cells or *Air-passages:* spaces in the tissue of leaves and some stems, p. 143.

Air-Plants, p. 34.

Akénium, or *akene.* See *achenium.*

Ala (plural *alæ*): a wing; the side-petals of a papilionaceous corolla, p. 105, fig. 218, *w.*

Alabástrum: a flower-bud.

Alar: situated in the forks of a stem.

Alate: winged, as the seeds of Trumpet-Creeper (fig. 316) the fruit of the Maple, Elm (fig. 301), &c.

Albescent: whitish, or turning white.

Absorption, p. 168.

Albúmen of the seed: nourishing matter stored up with the embryo, but not within it; p. 15, 136.

Albúmen, a vegetable product; a form of proteine, p. 165.

Albuminous (seeds): furnished with albumen, as the seeds of Indian corn (fig. 38, 39), of Buckwheat (fig 326), &c.

Albúrnum: young wood, sap-wood, p 153.

Alpine: belonging to high mountains above the limit of forests.

Alternate (leaves): one after another, p. 24, 71. Petals are *alternate with* the sepals, or stamens with the petals, when they stand over the intervals between them, p. 93.

Alveolate: honeycomb-like, as the receptacle of the Cotton-Thistle.

Ament: a catkin, p. 81. *Amentaceous:* catkin-like, or catkin-bearing.

Amorphous: shapeless; without any definite form.

Amphigástrium (plural *amphigastria*): a peculiar stipule-like leaf of certain Liverworts.

Amphitropous or *Amphitropal* ovules or seeds, p. 123, fig. 272.

Amplectant: embracing. *Amplexicaul* (leaves): clasping the stem by the base.

Ampullaceous: swelling out like a bottle or bladder.

Amyláceous: composed of starch, or starch-like.

Anántherous: without anthers. *Anánthous:* destitute of flowers ; flowerless.

Anástomosing: forming a net-work (*anastomosis*), as the veins of leaves.

Anátropous or *Anátropal* ovules or seeds ; p. 123, fig. 273.

Ancípital (*anceps*) : two-edged, as the stem of Blue-eyed Grass.

Andrœcium : a name for the stamens taken together.

Andrógynous : having both staminate and pistillate flowers in the same cluster or inflorescence, as many species of Carex.

Ándrophore : a column of united stamens, as in a Mallow ; or the support on which stamens are raised.

Anfráctuose : bent hither and thither, as the anthers of the Squash, &c.

Angiospérmœ, Angiospérmous Plants : with their seeds formed in an ovary or peri-carp, p. 183.

Angular divergence of leaves, p. 72.

Annual (plant) : flowering and fruiting the year it is raised from the seed, and then dying, p. 21.

Ánnular : in the form of a ring, or forming a circle.

Ánnulate : marked by rings ; or furnished with an

Ánnulus, or ring, like that of the spore-case of most Ferns (Manual Bot. N. States, plate 9, fig. 2) : in Mosses it is a ring of cells placed between the mouth of the spore-case and the lid, in many species.

Anterior, in the blossom, is the part next the bract, i. e. external : — while the posterior side is that next the axis of inflorescence. Thus, in the Pea, &c. the keel is *anterior,* and the standard *posterior.*

Anther : the essential part of the stamen, which contains the pollen ; p. 86, 113.

Antherídium (plural *antheridia*) : the organ in Mosses, &c. which answers to the anther of Flowering plants.

Antheríferous : anther-bearing.

Anthésis : the period or the act of the expansion of a flower.

Anthocárpous (fruits) : same as multiple fruits ; p. 133.

Ánticous : same as anterior.

Antrórse : directed upwards or forwards.

Apétalous : destitute of petals ; p. 90, fig. 179.

Aphýllous : destitute of leaves, at least of foliage.

Ápical : belonging to the apex or point.

Apículate : pointletted ; tipped with a short and abrupt point.

Apocárpous (pistils) : when the several pistils of the same flower are separate, as in a Buttercup, Sedum (fig. 168), &c.

Apóphysis : any irregular swelling ; the enlargement at the base of the spore-case of the Umbrella-Moss (Manual, plate 4), &c.

Appendage · any superadded part.

Appendículate : provided with appendages.

Appressed : where branches are close pressed to the stem, or leaves to the branch, &c.

Ápterous : wingless.

Aquatic : living or growing in water ; applied to plants whether growing under water, or with all but the base raised out of it.

Aráchnoid : cobwebby ; clothed with, or consisting of, soft downy fibres.

Arbóreous, Arborescent : tree-like, in size or form ; p. 36.

18

Archegónium (plural *archegonia*) : the organ in Mosses, &c., which is analogous to the pistil of Flowering Plants.

Árcuate: bent or curved like a bow.

Aréolate: marked out into little spaces or *areolæ.*

Árillate (seeds) · furnished with an

Aril or *Aríllus:* a fleshy growth forming a false coat or appendage to a seed; p. 135, fig. 318.

Arístate: awned, i. e. furnished with an arista, like the beard of Barley, &c.

Arístulate: diminutive of the last ; short-awned.

Arrow-shaped or *Arrow-headed:* same as *sagittate ;* p. 59, fig. 95.

Articulated: jointed ; furnished with joints or *articulations,* where it separates or inclines to do so. *Articulated leaves,* p. 64.

Artificial Classification, p. 196.

Ascending (stems, &c.), p. 37 ; (seeds or ovules), p. 122.

Aspergilliform : shaped like the brush used to sprinkle holy water; as the stigmas of many Grasses.

Assimilation, p. 162.

Assurgent: same as ascending, p. 37.

Átropous or *Átropal* (ovules) : same as orthotropous.

Auriculate: furnished with auricles or ear-like appendages, p. 59.

Awl-shaped: sharp-pointed from a broader base, p. 68.

Awn: the bristle or beard of Barley, Oats, &c. ; or any similar bristle-like appendage.

Awned: furnished with an awn or long bristle-shaped tip.

Axil: the angle on the upper side between a leaf and the stem, p. 20.

Axile: belonging to the axis, or occupying the axis ; p. 119, &c.

Axillary (buds, &c.) : occurring in an axil, p 21, 77, &c.

Axis: the central line of any body ; the organ round which others are attached ; the root and stem. *Ascending Axis,* p. 9. *Descending Axis,* p. 9.

Baccate: berry-like, of a pulpy nature like a berry (in Latin *bacca*) ; p. 127.

Barbate: bearded ; bearing tufts, spots, or lines of hairs.

Barbed: furnished with a *barb* or double hook ; as the apex of the bristle on the fruit of Echinospermum (Stickseed), &c.

Bárbellate: said of the bristles of the pappus of some Compositæ (species of Liatris, &c), when beset with short, stiff hairs, longer than when denticulate, but shorter than when plumose.

Barbéllulate: diminutive of barbellate.

Bark: the covering of a stem outside of the wood, p. 150, 152.

Basal: belonging or attached to the

Base: that extremity of any organ by which it is attached to its support.

Bast, Bast-fibres, p. 147.

Beaked: ending in a prolonged narrow tip.

Bearded: see *barbate. Beard* is sometimes used popularly for awn, more commonly for long or stiff hairs of any sort.

Bell-shaped : of the shape of a bell, as the corolla of Harebell, fig. 207, p. 102.

Berry: a fruit pulpy or juicy throughout, as a grape; p. 127.

Bi- (or *Bis*), in compound words : twice ; as

Biartículate: twice jointed, or two-jointed ; separating into two pieces.

Biauriculate: having two ears, as the leaf in fig. 96.

Bicallose: having two callosities or harder spots.

Bicdrinate: two-keeled, as the upper palea of Grasses.

Bicipital (*Biceps*) : two-headed ; dividing into two parts at the top or bottom.

Biconjugate: twice paired, as when a petiole forks twice.

Bidéntate: having two teeth (not twice or doubly dentate).

Biénnial: of two years' continuance ; springing from the seed one season, flowering and dying the next ; p. 21.

Bifárious: two-ranked ; arranged in two rows.

Bifid: two-cleft to about the middle, as the petals of Mouse-ear Chickweed.

Bifóliolate: a compound leaf of two leaflets ; p. 66.

Bifúrcate: twice forked ; or, more commonly, forked into two branches.

Bijugate: bearing two pairs (of leaflets, &c.).

Bilábiate: two-lipped, as the corolla of sage. &c , p. 105, fig. 209.

Bilamellate: of two plates (*lamellæ*), as the stigma of Mimulus.

Bilóbed: the same as two-lobed.

Bilócular: two-celled ; as most anthers, the pod of Foxglove, most Saxifrages (fig. 254), &c.

Binate: in couples, two together.

Bipartite: the Latin form of two-parted ; p. 62.

Bipinnate (leaf) : twice pinnate ; p. 66, fig. 130.

Bipinnátifid: twice pinnatifid, p. 64 ; that is, pinnatifid with the lobes again pinnatifid.

Biplícate: twice folded together.

Bisérial, or *Bisériate:* occupying two rows, one within the other.

Biserrate: doubly serrate, as when the teeth of a leaf, &c. are themselves serrate.

Bitérnate: twice ternate ; i. e. principal divisions 3, each bearing 3 leaflets, &c.

Bladdery: thin and inflated, like the calyx of Silene inflata.

Blade of a leaf: its expanded portion ; p 54.

Boat-shaped: concave within and keeled without, in shape like a small boat.

Bráchiate: with opposite branches at right angles to each other, as in the Maple and Lilac.

Bract (Latin, *bractea*). Bracts, in general, are the leaves of an inflorescence, more or less different from ordinary leaves. Specially, the bract is the small leaf or scale from the axil of which a flower or its pedicel proceeds ; p. 78 ; and a

Bractlet (*bracteola*) is a bract seated on the pedicel or flower-stalk ; p. 78, fig. 156.

Branch, p. 20, 36.

Bristles: stiff, sharp hairs, or any very slender bodies of similar appearance.

Bristly: beset with bristles.

Brush-shaped: see *aspergilliform.*

Bryology: that part of Botany which relates to Mosses.

Bud: a branch in its earliest or undeveloped state ; p 20.

Bud-scales, p. 22, 50.

Bulb: a leaf-bud with fleshy scales, usually subterranean ; p. 45, fig. 73.

Bulbíferous: bearing or producing bulbs.

Bulbose or *bulbous* : bulb-like in shape, &c.

Bulblets: small bulbs, borne above ground, as on the stems of the bulb-bearing Lily and on the fronds of Cistopteris bulbifera and some other Ferns; p. 46.

Bulb-scales, p. 50.

Bullate: appearing as if blistered or bladdery (from *bulla,* a bubble).

Caducous: dropping off very early, compared with other parts; as the calyx in the Poppy Family, falling when the flower opens.

Cæspitose, or *Céspitose:* growing in turf-like patches or tufts, like most sedges, &c.

Cálcarate: furnished with a spur (*calcar*), as the flower of Larkspur, fig. 183, and Violet, fig. 181.

Calcéolate or *Cálceiform:* slipper-shaped, like one petal of the Lady's Slipper.

Cállose: hardened ; or furnished with callosities or thickened spots.

Cálycine: belonging to the calyx.

Calýculate: furnished with an outer accessory calyx (*calyculus*) or set of bracts looking like a calyx, as in true Pinks.

Calýptra: the hood or veil of the capsule of a Moss: Manual, p. 607, &c.

Calýptriform: shaped like a calyptra or candle-extinguisher.

Calyx: the outer set of the floral envelopes or leaves of the flower ; p. 85.

Cambium and *Cambium-layer,* p. 154.

Campánulate: bell-shaped ; p. 102, fig. 207.

Campylótropous, or *Campylótropal;* curved ovules and seeds of a particular sort; p. 123, fig. 271.

Campylospérmous: applied to fruits of Umbelliferæ when the seed is curved in at the edges, forming a groove down the inner face ; as in Sweet Cicely.

Canalículate: channelled, or with a deep longitudinal groove.

Cáncellate: latticed, resembling lattice-work.

Canéscent: grayish-white ; hoary, usually because the surface is covered with fine white hairs. *Incanous* is whiter still.

Capilláceous, Cápillary: hair-like in shape ; as fine as hair or slender bristles.

Cápitate: having a globular apex, like the head on a pin ; as the stigma of Cherry, fig. 213; or forming a head, like the flower-cluster of Button-bush, fig. 161.

Capitéllate: diminutive of capitate; as the stigmas of fig. 255.

Capítulum (a little head): a close rounded dense cluster or head of sessile flowers; p. 80, fig. 161.

Capréolate: bearing tendrils (from *capreolus,* a tendril).

Capsule: a pod ; any dry dehiscent seed-vessel; p. 131, fig. 305, 306.

Cápsular: relating to, or like a capsule.

Carína: a keel ; the two anterior petals of a papilionaceous flower, which are combined to form a body shaped somewhat like the keel (or rather the prow) of a vessel ; p. 105, fig. 218, *k.*

Cárinate: keeled ; furnished with a sharp ridge or projection on the lower side.

Cariópsis, or *Caryópsis:* the one-seeded fruit or grain of Grasses, &c., p. 351.

Cárneous: flesh-colored ; pale red.

Cárnose: fleshy in texture.

Cárpel, or *Carpídium:* a simple pistil, or one of the parts or leaves of which a compound pistil is composed ; p. 117.

Cárpellary: pertaining to a carpel.

Carpology: that department of Botany which relates to fruits.

Cárpophore: the stalk or support of a fruit or pistil within the flower; as in fig. 276 – 278.

Curtilágínous, or *Cartilagíneous:* firm and tough, like cartilage, in texture.

Cáruncle: an excrescence at the scar of some seeds; as those of Polygala.

Carínculate: furnished with a caruncle.

Caryophylláceous: pink-like: applied to a corolla of 5 long-clawed petals; fig. 200.

Catkin: a scaly deciduous spike of flowers, an ament; p. 81.

Cuudate: tailed, or tail-pointed.

Cauder: a sort of trunk, such as that of Palms; an upright rootstock; p. 37.

Cauléscent: having an obvious stem; p. 36.

Cuúlicle: a little stem, or rudimentary stem; p 6.

Cuúline: of or belonging to a stem (*caulis,* in Latin), p. 36.

Cell (diminutive *Cellule*): the cavity of an anther, ovary, &c., p. 113, 119; one of the elements or vesicles of which plants are composed; p. 140, 142.

Céllular tissue of plants; p. 142. *Cellular Bark,* p. 152.

Cellulose, p. 159.

Centrífugal (inflorescence): produced or expanding in succession from the centre outwards; p. 82. The radicle is centrifugal, when it points away from the centre of the fruit.

Centrípetal: the opposite of centrifugal; p. 79, 83.

Cereal: belonging to corn, or corn-plants.

Cérnuous: nodding; the summit more or less inclining.

Chaff: small membranous scales or bracts on the receptacle of Compositæ; the glumes, &c. of Grasses.

Chaffy: furnished with chaff, or of the texture of chaff.

Chaldza: that part of the ovule where all the parts grow together; p. 122.

Channelled: hollowed out like a gutter; same as *canaliculate.*

Character: a phrase expressing the essential marks of a species, genus, &c. which distinguish it from all others; p. 180.

Chartáceous: of the texture of paper or parchment.

Chlórophyll: the green grains in the cells of the leaf, and of other parts exposed to the light, which give to herbage its green color; p. 155.

Chrómule: coloring matter in plants, especially when not green, or when liquid.

Cícatrix: the scar left by the fall of a leaf or other organ.

Cíliate: beset on the margin with a fringe of *cilia,* i. e. of hairs or bristles, like the eyelashes fringing the eyelids, whence the name.

Cinéreous, or *Cineráceous:* ash-grayish; of the color of ashes.

Círcinate: rolled inwards from the top, like a crosier, as the shoots of Ferns; p. 76, fig. 154; the flower-clusters of Heliotrope, &c.

Circumscissile, or *Circumcissile:* divided by a circular line round the sides, as the pods of Purslane, Plantain, &c.; p. 133, fig. 298, 311.

Circumscription: the general outline of a thing.

Cirrhíferous, or *Cirrhose:* furnished with a tendril (Latin, *cirrhus*); as the Grape-vine. *Cirrhose* also means resembling or coiling like tendrils, as the leaf-stalks of Virgin's-bower; p. 37.

Class, p 175, 177.

Classification, p. 173.

18 *

Cláthrate : latticed ; same as *cancellate.*

Clávate : club-shaped ; slender below and thickened upwards.

Claw: the narrow or stalk-like base of some petals, as of Pinks ; p. 102, fig 200.

Climbing : rising by clinging to other objects ; p. 37.

Club-shaped : see *clavate.*

Clustered : leaves, flowers, &c. aggregated or collected into a bunch

Clýpeate : buckler-shaped.

Coddunate : same as *connate ;* i. e. united.

Coaléscent : growing together.

Codrctate : contracted or brought close together.

Coated Bulbs, p 46.

Cobwebby : same as *arachnoid ;* bearing hairs like cobwebs or gossamer.

Cóccus (plural *cocci*): anciently a berry ; now mostly used to denote the carpels of a dry fruit which are separable from each other, as of Euphorbia.

Cochledriform : spoon-shaped.

Cóchleate : coiled or shaped like a snail-shell.

Cælospérmous : applied to those fruits of Umbelliferæ which have the seed hollowed on the inner face, by the curving inwards of the top and bottom ; as in Coriander.

Coherent, in Botany, is usually the same as *connate ;* p. 104.

Collective fruits, p. 133.

Collum or *Cóllar :* the neck or line of junction between the stem and the root.

Columélla : the axis to which the carpels of a compound pistil are often attached, as in Geranium (fig. 278), or which is left when a pod opens, as in Azalea and Rhododendron.

Column: the united stamens, as in Mallow, or the stamens and pistils united into one body, as in the Orchis family, fig. 226.

Columnar : shaped like a column or pillar.

Coma : a tuft of any sort (literally, a head of hair) ; p. 135, fig. 317.

Cómose : tufted ; bearing a tuft of hairs, as the seeds of Milkweed ; fig. 317.

Cómmissure : the line of junction of two carpels, as in the fruit of Umbelliferæ, such as Parsnip, Caraway, &c.

Common : used as "general," in contradistinction to "partial" ; e. g. "common involucre," p. 81.

Cómplanate : flattened.

Compound leaf, p. 64. *Compound pistil,* p. 118. *Compound umbel,* &c., p. 81.

Complete (flower), p. 89.

Complicate : folded upon itself.

Compressed : flattened on two opposite sides.

Condúplicate : folded upon itself lengthwise, as are the leaves of Magnolia in the bud, p. 76.

Cone : the fruit of the Pine family ; p. 133, fig. 314.

Cónfluent : blended together ; or the same as *coherent.*

Confôrmed : similar to another thing it is associated with or compared to ; or closely fitted to it, as the skin to the kernel of a seed.

Congésted, Conglómerate : crowded together.

Cónjugate : coupled ; in single pairs.

Connate : united or grown together from the first.

Connéctive, Connectívum : the part of the anther connecting its two cells ; p. 113.

Conntvent : converging, or brought close together.

Consolidated forms of vegetation, p. 47.

Continuous : the reverse of interrupted or articulated.

Contorted : twisted together. *Contorted æstivation :* same as *convolute;* p. 109.

Contortuplicate : twisted back upon itself.

Contracted : either narrowed or shortened.

Contrary : turned in an opposite direction to another organ or part with which it is compared.

Cónvolute : rolled up lengthwise, as the leaves of the Plum in vernation ; p. 76, fig. 151. In æstivation, same as *contorted;* p. 109.

Cordate : heart-shaped ; p. 58, fig. 90, 99.

Coriaceous : resembling leather in texture.

Corky : of the texture of cork. *Corky layer* of bark, p. 152.

Corm, Cornus : a solid bulb, like that of Crocus ; p. 44, fig. 71, 72.

Córneous : of the consistence or appearance of horn, as the albumen of the seed of the Date, Coffee, &c.

Corniculate : furnished with a small horn or spur.

Cornúte : horned ; bearing a horn-like projection or appendage.

Corólla : the leaves of the flower within the calyx ; p. 86.

Corolláceous, Corollíne : like or belonging to a corolla.

Coróna : a coronet or crown ; an appendage at the top of the claw of some petals, as Silene and Soapwort, fig. 200, or of the tube of the corolla of Hound's-Tongue, &c.

Coróuate : crowned ; furnished with a crown.

Córtical : belonging to the bark (*cortex*).

Córymb : a sort of flat or convex flower-cluster ; p. 79, fig. 158.

Corymbóse : approaching the form of a corymb, or branched in that way ; arranged in corymbs.

Costa : a rib ; the midrib of a leaf, &c. *Costate :* ribbed.

Cotylédous : the first leaves of the embryo ; p. 6, 137.

Cratériform : goblet-shaped ; broadly cup-shaped.

Creeping (stems) : growing flat on or beneath the ground and rooting; p. 37.

Crémocarp : a half-fruit, or one of the two carpels of Umbelliferæ.

Crenate, or *Crenelled :* the edge scalloped into rounded teeth ; p. 62, fig. 114.

Crested, or *Cristate :* bearing any elevated appendage like a crest.

Cribrose : pierced like a sieve with small apertures.

Crinite : bearded with long hairs, &c.

Crown : see *corona.*

Crowning : borne on the apex of anything.

Cruciate, or *Cruciform :* cross-shaped, as the four spreading petals of the Mustard (fig. 187), and all the flowers of that family.

Crustaceous : hard, and brittle in texture ; crust-like.

Cryptógamous, or *Cryptogamic :* relating to Cryptogamia ; p. 172, 197.

Cucúllate : hooded, or hood-shaped, rolled up like a cornet of paper, **or a hood** (*cucullus*), as the spathe of Indian Turnip, fig. 162.

Culm : a straw ; the stem of Grasses and Sedges.

Cúneate, Cúneiform : wedge-shaped ; p. 58, fig. 94.

Cup-shaped: same as cyathiform, or near it.

Cúpule: a little cup; the cup to the acorn of the Oak, p. 130, fig. 299.

Cúpulate: provided with a cupule.

Cúspidate: tipped with a sharp and stiff point.

Cut: same as incised, or applied generally to any sharp and deep division.

Cúticle: the skin of plants, or more strictly its external pellicle.

Cyáthiform: in the shape of a cup, or particularly of a wine-glass.

Cýcle: one complete turn of a spire, or a circle; p. 73.

Cýclical: rolled up circularly, or coiled into a complete circle.

Cyclósis: the circulation in closed cells, p. 167.

Cylindraceous: approaching to the

Cylíndrical form; as that of stems, &c., which are round, and gradually if at all tapering.

Cýmbæform, or *Cymbiform:* same as boat-shaped.

Cyme: a cluster of centrifugal inflorescence, p 82, fig. 165, 167.

Cýmose: furnished with cymes, or like a cyme.

Deca- (in composition of words of Greek derivation) : ten; as

Decágynous: with 10 pistils or styles. *Decándrous:* with 10 stamens.

Decíduous: falling off, or subject to fall, said of leaves which fall in autumn, and of a calyx and corolla which fall before the fruit forms.

Declined: turned to one side, or downwards, as the stamens of Azalea nudiflora.

Decompound: several times compounded or divided; p 67, fig. 138.

Decumbent: reclined on the ground, the summit tending to rise; p. 37.

Decurrent (leaves): prolonged on the stem beneath the insertion, as in Thistles.

Decússate: arranged in pairs which successively cross each other; fig. 147.

Definite: when of a uniform number, and not above twelve or so.

Deflexed: bent downwards.

Deflorate: past the flowering state, as an anther after it has discharged its pollen.

Dehiscence: the mode in which an anther or a pod regularly bursts or splits open; p. 132.

Dehiscent: opening by regular dehiscence.

Deliquescent: branching off so that the stem is lost in the branches, p. 25.

Deltoid: of a triangular shape, like the Greek capital Δ.

Demersed: growing below the surface of water.

Dendroid, Dendritic: tree-like in form or appearance.

Dentate: toothed (from the Latin *dens,* a tooth), p. 61, fig. 113.

Denticulate: furnished with denticulations, or very small teeth : diminutive of the last.

Depauperate (impoverished or starved): below the natural size.

Depressed: flattened, or as if pressed down from above; flattened vertically.

Descending: tending gradually downwards.

Determinate Inflorescence, p. 81, 83.

Dextrorse: turned to the right hand.

Di- (in Greek compounds): two, as

Diádelphous (stamens): united by their filaments in two sets; p. 111, fig. 227.

Diándrous: having two stamens, p. 112.

Diagnosis: a short distinguishing character, or descriptive phrase.

Diáphanous: transparent or translucent.

Dichlamýdeous (flower) : having both calyx and corolla.

Dichótomons: two-forked.

Diclínous: having the stamens in one flower, the pistils in another; p. 89, fig. 176, 177.

Dicóccous (fruit) : splitting into two *cocci,* or closed carpels.

Dicotylédonous (embryo) : having a pair of cotyledons ; p. 16, 137.

Dicotyledonous Plants, p. 150, 182.

Didymous: twin.

Didýnamous (stamens) ; having four stamens in two pairs, one pair shorter than the other, as in fig. 194, 195.

Diffuse: spreading widely and irregularly.

Digitate (fingered) : where the leaflets of a compound leaf are all borne on the apex of the petiole ; p. 65, fig. 129.

Digynous (flower) : having two pistils or styles, p. 116.

Dímerous: made up of two parts, or its organs in twos.

Dimidiate: halved ; as where a leaf or leaflet has only one side developed, or a stamen has only one lobe or cell ; fig. 239.

Dimorphous: of two forms.

Dicecious, or *Dioícous:* where the stamens and pistils are in separate flowers on different plants ; p. 89.

Dipétalous: of two petals. *Diphýllous:* two-leaved. *Dípterous:* two-winged.

Disciform or *Disk-shaped:* flat and circular, like a disk or quoit.

Disk: the face of any flat body ; the central part of a head of flowers, like the Sunflower, or Coreopsis (fig. 224), as opposed to the *ray* or margin; a fleshy expansion of the receptacle of a flower ; p. 125.

Dissected: cut deeply into many lobes or divisions.

Dissépiments: the partitions of an ovary or a fruit ; p. 119.

Dístichous: two-ranked ; p. 73.

Distinct: uncombined with each other ; p. 102.

Divaricate: straddling ; very widely divergent.

Divided (leaves, &c.) : cut into divisions extending about to the base or the mid rib; p. 62, fig. 125.

Dodeca- (in Greek compounds) : twelve ; as

Dodecágynous: with twelve pistils or styles.

Dodecandrous: with twelve stamens.

Dolabriform: axe-shaped.

Dorsal: pertaining to the back (*dorsum*) of an organ.

Dorsal Suture, p. 117.

Dotted Ducts, p. 148.

Double Flowers, so called : where the petals are multiplied unduly ; p. 85, 98.

Downy: clothed with a coat of soft and short hairs.

Drupe: a stone-fruit ; p. 128, fig. 285.

Drupaceous: like or pertaining to a drupe.

Ducts: the so-called vessels of plants ; p. 146, 148.

Dumose: bushy, or relating to bushes.

Duramen: the heart-wood, p. 153.

Dwarf: remarkably low in stature.

E-, or *Ex-*, at the beginning of compound words, means destitute of; as *ecostate*, without a rib or midrib; *exalbuminous*, without albumen, &c.

Eared: see *auriculate;* p. 59, fig. 96.

Ebrácteate; destitute of bracts.

Echínate: armed with prickles (like a hedgehog). *Echínulate:* a diminutive of it.

Edentate: toothless.

Effete: past bearing, &c.; said of anthers which have discharged their pollen.

Eglandulose: destitute of glands.

Eláters: threads mixed with the spores of Liverworts. (Manual, p. 682.)

Ellipsoidal: approaching an elliptical figure.

Elliptical: oval or oblong, with the ends regularly rounded; p. 58, fig. 88.

Emárginate: notched at the summit; p. 60, fig. 108.

Émbryo: the rudimentary undeveloped plantlet in a seed; p. 6, fig. 9, 12, 26, 31–37, &c., and p. 136. *Embryo-sac,* p. 139.

Emersed: raised out of water.

Endecágynous: with eleven pistils or styles. *Endecándrous:* with eleven stamens.

Éndocarp: the inner layer of a pericarp or fruit; p. 128.

Éndochrome: the coloring matter of Algæ and the like.

Endógenous Stems, p. 150. *Endogenous Plants,* p. 150.

Endosmose: p. 168.

Éndosperm: another name for the albumen of a seed.

Éndostome: the orifice in the inner coat of an ovule.

Eunea-: nine. *Enneágynous:* with nine petals or styles.

Enneándrous: with nine stamens.

Ensiform: sword-shaped; as the leaves of Iris, fig. 134.

Entire: the margins not at all toothed, notched, or divided, but even; p. 61.

Ephemeral: lasting for a day or less, as the corolla of Purslane, &c.

Epi-, in composition: upon; as

Épicarp: the outermost layer of a fruit; p. 128.

Epidermal: relating to the *Epidérmis,* or the skin of a plant; p. 152, 155.

Epigæous: growing on the earth, or close to the ground.

Epígynous: upon the ovary; p. 105, 111.

Epipétalous: borne on the petals or the corolla.

Epiphýllous: borne on a leaf.

Epiphyte: a plant growing on another plant, but not nourished by it; p. 34.

Epiphýtic or *Epiphýtal:* relating to *Epiphytes;* p. 34.

Epísperm: the skin or coat of a seed, especially the outer coat.

Equal: same as *regular;* or of the same number or length, as the case may be, of the body it is compared with.

Equally pinnate: same as abruptly pinnate; p. 65.

Équitant (riding straddle); p. 68, fig. 133, 134.

Erose: eroded, as if gnawed.

Eróstrate: not beaked.

Essential Organs of the flower, p 85.

Estivátion: see *æstivation.*

Etiolated: blanched by excluding the light, as the stalks of Celery.

Evergreen: holding the leaves over winter and until new ones appear, or longer.

Exalbuminous (seed): destitute of albumen; p. 136.

Excurrent: running out, as when a midrib projects beyond the apex of a leaf, or a trunk is continued to the very top of a tree.

Exhalation, p. 156, 169.

Exógenous Stems, p. 150. *Exogenous Plants*, p. 182.

Exostome: the orifice in the outer coat of the ovule ; p. 122.

Explanate: spread or flattened out.

Exserted: protruding out of, as the stamens out of the corolla of fig. 201.

Exstípulate: destitute of stipules.

Extra-axillary: said of a branch or bud a little out of the axil ; as the upper accessory buds of the Butternut, p 27, fig. 52.

Extrórse: turned outwards ; the anther is extrorse when fastened to the filament on the side next the pistil, and opening on the outer side, as in Iris ; p. 113.

Falcate: scythe-shaped ; a flat body curved, its edges parallel.

Family: p. 176

Farinaceous: mealy in texture. *Fárinose*: covered with a mealy powder.

Fásciate: banded ; also applied to monstrous stems which grow flat.

Fáscicle: a close cluster ; p. 83.

Fáscicled, Fascículated: growing in a bundle or tuft, as the leaves of Pine and Larch (fig 139, 140), the roots of Pæony and Dahlia, fig. 60.

Fastigiate: close, parallel, and upright, as the branches of Lombardy Poplar.

Faux (plural, *fauces*) : the throat of a calyx, corolla, &c.

Favéolate, Fávose: honeycombed ; same as *alveolate*

Feather-veined: where the veins of a leaf spring from along the sides of a mid rib ; p. 57, fig. 86 – 94.

Female (flowers) : with pistils and no stamens.

Fenéstrate: pierced with one or more large holes, like windows.

Ferrugineous, or *Ferruginous*: resembling iron-rust ; red-grayish.

Fertile: fruit-bearing, or capable of producing fruit; also said of anthers when they produce good pollen.

Fertilization: the process by which pollen causes the embryo to be formed.

Fibre, p. 145. *Fibrous*: containing much fibre, or composed of fibres.

Fibrillose: formed of small fibres.

Fibrine, p. 165.

Fiddle-shaped: obovate with a deep recess on each side.

Filament: the stalk of a stamen ; p. 86, fig. 170, *a*; also any slender thread-shaped appendage.

Filaméntose, or *Filamentous*: bearing or formed of slender threads.

Filiform: thread-shaped ; long, slender, and cylindrical.

Fimbriate: fringed ; furnished with fringes (*fimbriæ*).

Fistular or *Fistulose*: hollow and cylindrical, as the leaves of the Onion.

Flabélliform or *Flabéllate*: fan-shaped ; broad, rounded at the summit, and narrowed at the base.

Flágellate, or *Flagélliform*: long, narrow, and flexible, like the thong of a whip or like the runners (*flagellæ*) of the Strawberry.

Flavescent: yellowish, or turning yellow.

Fleshy: composed of firm pulp or flesh.

Fleshy Plants, p. 47.

Fléxuose, or *Fléxuous:* bending gently in opposite directions, in a zigzag way.

Floating: swimming on the surface of water.

Flóccose: composed, or bearing tufts, of woolly or long and soft hairs.

Flora (the goddess of flowers): the plants of a country or district, taken together, or a work systematically describing them ; p. 3.

Floral: relating to the blossom

Floral Envelopes: the leaves of the flower; p. 85, 99.

Floret: a diminutive flower ; one of the flowers of a head (or of the so-called compound flower) of Compositæ, p. 106.

Flower: the whole organs of reproduction of Phænogamous plants; p. 84.

Flower-bud: an unopened flower.

Flowering Plants, p. 177. *Flowerless Plants*, p. 172, 177.

Foliáceous: belonging to, or of the texture or nature of, a leaf (*folium*).

Fóliose: leafy; abounding in leaves.

Fóliolate: relating to or bearing leaflets (*foliola*).

Fóllicle: a simple pod, opening down the inner suture ; p. 131, fig. 302.

Follicular: resembling or belonging to a follicle.

Food of Plants, p. 160.

Foramen: a hole or orifice, as that of the ovule ; p. 122.

Fornix: little arched scales in the throat of some corollas, as of Comfrey.

Fórnicate: over-arched, or arching over.

Fóveate: deeply pitted. *Fovéolate:* diminutive of *foveate*.

Free: not united with any other parts of a different sort ; p. 103.

Fringed: the margin beset with slender appendages, bristles, &c.

Frond: what answers to leaves in Ferns ; the stem and leaves fused into one body, as in Duckweed and many Liverworts, &c.

Frondescence: the bursting into leaf.

Fróndose: frond-bearing ; like a frond : or sometimes used for leafy.

Fructification: the state of fruiting. Organs of, p. 76.

Fruit: the matured ovary and all it contains or is connected with ; p. 126.

Frutéscent: somewhat shrubby ; becoming a shrub (*frutex*).

Frutículose: like a small shrub. *Fruticose:* shrubby ; p. 36.

Fugacious: soon falling off or perishing.

Fulvous: tawny ; dull yellow with gray.

Funículus: the stalk of a seed or ovule; p. 122.

Funnel-form, or *Funnel-shaped:* expanding gradually upwards, like a funnel or tunnel ; p. 102.

Fúrcate: forked.

Furfuráceous: covered with bran-like fine scurf.

Furrowed: marked by longitudinal channels or grooves.

Fuscous: deep gray-brown.

Fúsiform: spindle-shaped ; p. 32.

Gáleate: shaped like a helmet (*galea*) ; as the upper sepal of the Monkshood, fig. 185, and the upper lip of the corolla of Dead-Nettle, fig. 209.

Gamopétalous: of united petals ; same as *monopetalous*, and a better word ; p. 102.

Gamophyllous: formed of united leaves. *Gamosépalous:* formed of united sepals.

Gelatine, p. 165.

Géminate: twin ; in pairs ; as the flowers of Linnæa.

Gemma : a bud.

Gemmation : the state of budding, or the arrangement of parts in the bud. ⸱

Gémmule : a small bud ; the buds of Mosses ; the plumule, p. 6.

Genículate : bent abruptly, like a knee (*genu*), as many stems.

Genus : a kind ; a rank above species ; p. 175, 176.

Generic Names, p. 178. *Generic Character,* p. 181.

Geographical Botany : the study of plants in their geographical relations, p. 3.

Germ : a growing point ; a young bud ; sometimes the same as embryo ; p. 136.

Germen : the old name for ovary.

Germination : the development of a plantlet from the seed ; p. 5, 137.

Gibbous : more tumid at one place or on one side than the other.

Glabrate : becoming glabrous with age, or almost glabrous.

Glabrous : smooth, i. e. having no hairs, bristles, or other pubescence.

Gladiate : sword-shaped ; as the leaves of Iris, fig. 134.

Glands : small cellular organs which secrete oily or aromatic or other products : they are sometimes sunk in the leaves or rind, as in the Orange, Prickly Ash, &c ; sometimes on the surface as small projections ; sometimes raised on hairs or bristles (*glandular hairs, &c.*), as in the Sweetbrier and Sundew. The name is also given to any small swellings, &c., whether they secrete anything or not.

Glandular, Glandulose : furnished with glands, or gland-like.

Glans (Gland) : the acorn or mast of Oak and similar fruits.

Glaucescent : slightly glaucous, or bluish-gray.

Glaucous : covered with a *bloom,* viz. with a fine white powder that rubs off, like that on a fresh plum, or a cabbage-leaf.

Globose : spherical in form, or nearly so. *Globular :* nearly globose.

Glochídiate (hairs or bristles) : barbed ; tipped with barbs, or with a double hooked point.

Glómerate : closely aggregated into a dense cluster.

Glómerule : a dense head-like cluster ; p. 83.

Glossology : the department of Botany in which technical terms are explained.

Glumaceous : glume-like, or glume-bearing.

Glume : Glumes are the husks or floral coverings of Grasses, or, particularly, the outer husks or bracts of each spikelet. (Manual, p. 535)

Glumelles : the inner husks, or paleæ, of Grasses.

Gluten : a vegetable product containing nitrogen ; p. 165.

Granular : composed of grains. *Granule :* a small grain.

Growth, p 138.

Grumous or *Grumose :* formed of coarse clustered grains.

Guttate : spotted, as if by drops of something colored.

Gymnocárpous : naked-fruited.

Gymnospérmous : naked-seeded ; p. 121.

Gymnospérmæ, or *Gymnospermous Plants,* p. 184 ; Manual, p. xxiii.

Gyndndrous : with stamens borne on, i. e. united with, the pistil ; p. 111, fig. 226.

Gynœcium : a name for the pistils of a flower taken altogether.

Gynobase : a particular receptacle or support of the pistils, or of the carpels of a compound ovary, as in Geranium, fig. 277, 278.

Gynophore: a stalk raising a pistil above the stamens, as in the Cleome Family, p. 276.

Gyrate: coiled in a circle : same as *circinate.*

Gyrose: strongly bent to and fro.

Habit : the general aspect of a plant, or its mode of growth.

Habitat : the situation in which a plant grows in a wild state.

Hairs: hair-like projections or appendages of the surface of plants.

Hairy : beset with hairs, especially longish ones.

Halberd-shaped, or *Halberd-headed:* see *hastate.*

Halved: when appearing as if one half of the body were cut away.

Hamate or *Hamose:* hooked ; the end of a slender body bent round.

Hamulose: bearing a small hook ; a diminutive of the last.

Hastate or *Hastile:* shaped like a halberd ; furnished with a spreading lobe on each side at the base ; p. 59, fig. 97.

Heart-shaped : of the shape of a heart as commonly painted ; p. 58, fig. 90.

Heart-wood : the older or matured wood of exogenous trees ; p. 153.

Helicoid : coiled like a *helix* or snail-shell.

Helmet : the upper sepal of Monkshood in this shape, fig. 185, &c.

Hemi- (in compounds from the Greek): half; e. g. *Hemispherical,* &c.

Hemicarp: half-fruit, or one carpel of an Umbelliferous plant.

Hemitropous or *Hemitropal* (ovule or seed): nearly same as *amphitropous,* p. 123.

Hepta- (in words of Greek origin): seven; as,

Heptagynous: with seven pistils or styles.

Heptamerous: its parts in sevens. *Heptandrous:* having seven stamens.

Herb, p. 20.

Herbaceous: of the texture of common herbage ; not woody ; p. 36.

Herbarium: the botanist's arranged collection of dried plants ; p. 201.

Hermaphrodite (flower): having both stamens and pistils in the same blossom ; same as *perfect;* p. 89.

Heterocarpous: bearing fruit of two sorts or shapes, as in Amphicarpæa.

Heterogamous: bearing two or more sorts of flowers as to their stamens and pistils ; as in Aster, Daisy, and Coreopsis.

Heteromorphous : of two or more shapes.

Heterotropous, or *Heterotropal* (ovule) : the same as *amphitropous;* p. 123.

Hexa- (in Greek compounds): six ; as

Hexagonal: six-angled. *Hexagynous:* with six pistils or styles.

Hexamerous: its parts in sixes. *Hexandrous:* with six stamens.

Hexapterous: six-winged.

Hilar: belonging to the hilum.

Hilum: the scar of the seed ; its place of attachment ; p. 122, 135.

Hippocrepiform: horseshoe-shaped.

Hirsute: hairy with stiffish or beard-like hairs.

Hispid: bristly ; beset with stiff hairs. *Hispidulous* is a diminutive of it.

Hoary : grayish-white ; see *canescent,* &c.

Homogamous: a head or cluster with flowers all of one kind, as in Eupatorium.

Homogeneous: uniform in nature ; all of one kind.

Homomallous (leaves, &c.): originating all round a stem, but all bent or curved round to one side.

Homomórphous: all of one shape.
Homótropous or *Homótropal* (embryo): curved with the seed; curved one way.
Hood: same as *helmet* or *galea. Hooded:* hood-shaped ; see *cucullate.*
Hooked: same as *hamate.*
Horn: a spur or some similar appendage. *Horny:* of the texture of horn.
Hortus Siccus: an herbarium, or collection of dried plants ; p. 201.
Humifuse: spread over the surface of the ground.
Hyaline: transparent, or partly so.
Hybrid: a cross-breed between two allied species.
Hypocratériform: salver-shaped ; p. 101, fig. 202, 208.
Hypogéan: produced under ground.
Hypógynous: inserted under the pistil ; p. 103, fig. 212.

Icosándrous: having 12 or more stamens inserted on the calyx.
Ímbricate, Imbricated, Imbricative: overlapping one another, like tiles or shingles
   on a roof, as the scales of the involucre of Zinnia, &c., or the bud-scales of
   Horsechesnut (fig. 48) and Hickory (fig. 49). In æstivation, where some
   leaves of the calyx or corolla are overlapped on both sides by others ; p. 109.
Immarginate: destitute of a rim or border.
Immersed: growing wholly under water.
Impari-pinnate: pinnate with a single leaflet at the apex ; p. 65, fig. 126.
Imperfect flowers: wanting either stamens or pistils ; p. 89.
Inæquilateral: unequal-sided, as the leaf of a Begonia.
Incanous: hoary with white pubescence.
Incised: cut rather deeply and irregularly ; p. 62.
Included: enclosed ; when the part in question does not project beyond another. .
Incomplete Flower: wanting calyx or corolla ; p. 90.
Incrassated: thickened.
Incumbent: leaning or resting upon : the cotyledons are incumbent when the
   back of one of them lies against the radicle; the anthers are incumbent
   when turned or looking inwards, p. 113.
Incurved: gradually curving inwards.
Indefinite: not uniform in number, or too numerous to mention (over 12).
Indefinite or *Indeterminate Inflorescence:* p. 77.
Indehiscent: not splitting open ; i. e. not dehiscent; p. 127.
Indígenous: native to the country.
Individuals: p. 173.
Indúplicate: with the edges turned inwards ; p. 109.
Indúsium: the shield or covering of a fruit-dot of a Fern. (Manual, p 588.)
Inferior: growing below some other organ ; p. 104, 121.
Inflated: turgid and bladdery.
Inflexed: bent inwards.
Inflorescence: the arrangement of flowers on the stem ; p. 76.
Infra-axillary: situated beneath the axil.
Infundíbuliform or *Infundíbular:* funnel-shaped ; p. 102, fig. 199.
Innate (anther): attached by its base to the very apex of the filament; p. 113.
Innovation: an incomplete young shoot, especially in Mosses.
Inorganic Constituents, p. 160.

Insertion: the place or the mode of attachment of an organ to its support; p. 72.

Intercellular Passages or *Spaces*, p. 143, fig. 341.

Internode: the part of a stem between two nodes; p. 42.

Interruptedly pinnate: pinnate with small leaflets intermixed with larger ones, as in Water Avens.

Intrafoliaceous (stipules, &c.) : placed between the leaf or petiole and the stem.

Introrse: turned or facing inwards, i. e. towards the axis of the flower; p. 113.

Inverse or *Inverted:* where the apex is in the direction opposite to that of the organ it is compared with.

Involucel: a partial or small involucre; p. 81.

Involucellate: furnished with an involucel.

Involucrate: furnished with an involucre.

Involucre: a whorl or set of bracts around a flower, umbel, or head; p. 79.

Involute, in vernation, p. 76 : rolled inwards from the edges.

Irregular Flowers, p. 91.

Jointed: separate or separable at one or more places into pieces; p. 64, &c.

Keel: a projecting ridge on a surface, like the keel of a boat; the two anterior petals of a papilionaceous corolla; p. 105, fig. 217, 218, *k.*

Keeled: furnished with a keel or sharp longitudinal ridge.

Kernel of the ovule and seed, p. 122, 136.

Kidney-shaped: resembling the outline of a kidney; p. 59, fig. 100.

Labellum: the odd petal in the Orchis Family.

Labiate: same as *bilabiate* or two-lipped; p. 105.

Laciniate: slashed; cut into deep narrow lobes (called *laciniæ*).

Lactescent: producing milky juice, as does the Milkweed, &c.

Lacunose: full of holes or gaps.

Lævigate: smooth as if polished.

Lamellar or *Lamellate:* consisting of flat plates (*lamellæ*).

Lamina: a plate or blade : the blade of a leaf, &c., p 54.

Lanate: woolly; clothed with long and soft entangled hairs.

Lanceolate: lance-shaped; p. 58, fig. 86.

Lanuginous: cottony or woolly.

Latent buds: concealed or undeveloped buds; p. 26, 27.

Lateral: belonging to the side.

Latex: the milky juice, &c. of plants.

Lax: loose in texture, or sparse; the opposite of crowded.

Leaf, p. 49. *Leaf-buds,* p. 20, 27.

Leaflet: one of the divisions or blades of a compound leaf; p. 64.

Leaf-like: same as *foliaceous.*

Leathery: of about the consistence of leather; coriaceous.

Legume: a simple pod, dehiscent into two pieces, like that of the Pea, p. 131, fig. 303; the fruit of the Pea Family (*Leguminosæ*), of whatever shape.

Legumine, p. 165.

Leguminous: belonging to legumes, or to the Leguminous Family.

Lenticular: lens-shaped; i. e. flattish and convex on both sides.

Lépidote : leprous ; covered with scurfy scales.

Liber : the inner, fibrous bark of Exogenous plants ; p. 152.

Ligneous, or *Lignose :* woody in texture.

Ligulate : furnished with a ligule ; p. 106.

Ligule : the strap-shaped corolla in many Compositæ, p. 106, fig. 220 ; the little membranous appendage at the summit of the leaf-sheaths of most Grasses.

Limb : the blade of a leaf, petal, &c. ; p. 54, 102.

Linear : narrow and flat, the margins parallel ; p. 58, fig. 85.

Lineate : marked with parallel lines. *Lineolate :* marked with minute lines.

Lingulate, Linguiform : tongue-shaped.

Lip : the principal lobes of a bilabiate corolla or calyx, p. 105 ; the odd and peculiar petal in the Orchis Family.

Lobe : any projection or division (especially a rounded one) of a leaf, &c.

Locéllus (plural *locelli*) : a small cell, or compartment of a cell, of an ovary or anther.

Lócular : relating to the cell or compartment (*loculus*) of an ovary, &c.

Loculicídal (dehiscence) : splitting down through the middle of the back of each cell ; p. 132, fig 305.

Locústa : a name for the spikelet of Grasses.

Lóment : a pod which separates transversely into joints ; p. 131, fig. 304.

Lomentáceous : pertaining to or resembling a loment.

Lórate : thong-shaped.

Lúnate : crescent-shaped. *Lunulate :* diminutive of *lunate.*

Lýrate : lyre-shaped ; a pinnatifid leaf of an obovate or spatulate outline, the end-lobe large and roundish, and the lower lobes small, as in Winter-Cress and Radish, fig. 59.

Mace : the aril of the Nutmeg ; p. 135.

Mdculate : spotted or blotched.

Male (flowers) : having stamens but no pistil.

Mdmmose : breast-shaped.

Marcescent : withering without falling off.

Marginal : belonging to the edge or margin.

Marginate : margined, with an edge different from the rest.

Masked : see *personate.*

Median : belonging to the middle.

Medúllary : belonging to, or of the nature of pith (*medulla*) ; pithy.

Medullary Rays : the silver-grain of wood ; p. 151.

Medullary Sheath : a set of ducts just around the pith ; p. 151.

Membranaceous or *Mémbranous :* of the texture of membrane ; thin and more or less translucent.

Meniscoid : crescent-shaped.

Méricarp : one carpel of the fruit of an Umbelliferous plant.

Merismatic : separating into parts by the formation of partitions within.

Mésocarp : the middle part of a pericarp, when that is distinguishable into three layers ; p. 128.

Mesophlæum : the middle or green bark.

Micropyle : the closed orifice of the seed ; p. 135.
Midrib : the middle or main rib of a leaf; p. 55.
Milk-Vessels : p. 148.
Miniate : vermilion-colored.
Mitriform : mitre-shaped ; in the form of a peaked cap.
Monadélphous : stamens united by their filaments into one set; p. 111.
Monándrous (flower): having only one stamen; p. 112.
Monlíform : necklace-shaped ; a cylindrical body contracted at intervals.
Monochlamýdeous : having only one floral envelope, i. e. calyx but no corolla, as
    Anemone, fig. 179, and Castor-oil Plant, fig. 178.
Monocotylédonous (embryo): with only one cotyledon ; p. 16, 137.
Monocotyledonous Plants, p. 150, 192.
Monæcious, or *Monoicous* (flower): having stamens or pistils only ; p. 90.
Monógynous (flower) : having only one pistil, or one style; p. 116.
Monopétalous (flower) : with the corolla of one piece; p. 101.
Monophýllous : one-leaved, or of one piece ; p. 102.
Monosépalous : a calyx of one piece; i. e. with the sepals united into one body ;
    p. 101.
Monospérmous : one-seeded.
Monstrosity : an unnatural deviation from the usual structure or form.
Morphology : the department of botany which treats of the forms which an organ
    (say a leaf) may assume ; p. 28.
Múcronate : tipped with an abrupt short point (*mucro*) ; p. 60, fig. 111.
Mucrónulate : tipped with a minute abrupt point ; a diminutive of the last.
Multi-, in composition : many ; as
Multangular : many-angled. *Multicípital :* many-headed, &c.
Multifarious : in many rows or ranks. *Múltifid :* many-cleft ; p. 62.
Multilócular : many-celled. *Multisérial :* in many rows.
Multiple Fruits, p. 133.
Múricate : beset with short and hard points.
Múriform : wall-like ; resembling courses of bricks in a wall.
Muscology : the part of descriptive botany which treats of Mosses (i. e. *Musci*).
Múticous : pointless ; beardless ; unarmed.
Mycélium : the spawn of Fungi ; i. e. the filaments from which Mushrooms, &c.
    originate.

Nápiform : turnip-shaped ; p. 31, fig. 57.
Natural System : p. 195.
Naturalized : introduced from a foreign country, but growing perfectly wild and
    propagating freely by seed.
Navícular : boat-shaped, like the glumes of most Grasses.
Necklace-shaped : looking like a string of beads ; see *moniliform.*
Nectar : the honey, &c. secreted by glands, or by any part of the corolla.
Nectaríferous : honey-bearing ; or having a nectary.
Nectary : the old name for petals and other parts of the flower when of unusual
    shape, especially when honey-bearing. So the hollow spur-shaped petals of
    Columbine were called nectaries ; also the curious long-clawed petals of
    Monkshood, fig. 186, &c.

Needle-shaped: long, slender, and rigid, like the leaves of Pines; p. 68, fig. 140.

Nerve: a name for the ribs or veins of leaves, when simple and parallel; p. 56.

Nerved: furnished with nerves, or simple and parallel ribs or veins; p. 56, fig. 84.

Netted-veined: furnished with branching veins forming network; p. 56, fig. 83.

Nodding (in Latin form, *Nutant*): bending so that the summit hangs downward.

Node: a knot; the "joints" of a stem, or the part whence a leaf or a pair of leaves springs; p. 40.

Nodose: knotty or knobby. *Nodulose:* furnished with little knobs or knots.

Normal: according to rule; the pattern or natural way according to some law.

Notate: marked with spots or lines of a different color.

Nucumentaceous: relating to or resembling a small nut.

Nuciform: nut-shaped or nut-like. *Nucule:* a small nut.

Nucleus: the kernel of an ovule (p. 122) or seed (p. 136) of a cell; p. 140.

Nut: a hard, mostly one-seeded indehiscent fruit; as a chestnut, butternut, acorn; p. 130, fig. 299.

Nutlet: a little nut; or the stone of a drupe.

Ob- (meaning over against): when prefixed to words, signifies inversion; as,

Obcompressed: flattened the opposite of the usual way.

Obcordate: heart-shaped with the broad and notched end at the apex instead of the base; p. 60, fig. 109.

Oblanceolate: lance-shaped with the tapering point downwards; p. 58, fig. 91.

Oblique: applied to leaves, &c. means unequal-sided.

Oblong: from two to four times as long as broad, and more or less elliptical in outline; p. 58, fig. 87.

Obovate: inversely ovate, the broad end upward; p. 58, fig. 93.

Obtuse: blunt, or round at the end; p. 60, fig. 105.

Obverse: same as *inverse.*

Obvolute (in the bud): when the margins of one leaf alternately overlap those of the opposite one.

Ochreate: furnished with *ochreæ* (boots), or stipules in the form of sheaths; as in Polygonum, p. 69, fig. 137.

Ochroleucous: yellowish-white; dull cream-color.

Octo-, eight, enters into the composition of

Octagynous: with eight pistils or styles.

Octamerous: its parts in eights. *Octandrous:* with eight stamens, &c.

Offset: short branches next the ground which take root; p. 38.

One-ribbed, One-nerved, &c.: furnished with only a single rib, &c., &c.

Opaque, applied to a surface, means dull, not shining.

Operculate: furnished with a lid or cover (*operculum*), as the capsules of Mosses.

Opposite: said of leaves and branches when on opposite sides of the stem from each other (i. e. in pairs); p. 23, 71. Stamens are opposite the petals, &c. when they stand before them.

Orbicular, Orbiculate: circular in outline or nearly so; p. 58.

Organ: any member of the plant, as a leaf, a stamen, &c.; p. 1.

Organs of Vegetation, p. 7; of *Reproduction,* p. 77.

Organized, Organic: p. 1, 158, 159, 162.

Organic Constituents, p. 160. *Organic Structure,* p. 142.

Orthótropous or *Orthótropal* (ovule or seed) : p. 122, 135, fig. 270, 274.
Osseous: of a bony texture.
Oval: broadly elliptical ; p. 88.
Ovary: that part of the pistil containing the ovules or future seeds ; p. 86, 116.
Ovate: shaped like an egg with the broader end downwards, or, in plane sur-
    faces, such as leaves, like the section of an egg lengthwise ; p. 58, fig. 89.
Ovoid: ovate or oval in a solid form.
Ovule: the body which is destined to become a seed ; p. 86, 116, 122.

Palea (plural *paleæ*) : chaff ; the inner husks of Grasses ; the chaff or bracts on
    the receptacle of many Compositæ, as Coreopsis, fig. 220, and Sunflower.
Paleaceous: furnished with chaff, or chaffy in texture.
Palmate: when leaflets or the divisions of a leaf all spread from the apex of the
    petiole, like the hand with the outspread fingers ; p. 167, fig. 129, &c.
Palmately (veined, lobed, &c.) : in a palmate manner ; p. 57, 63, 65.
Pandúriform: fiddle-shaped (which see).
Pánicle: an open cluster ; like a raceme, but more or less compound ; p. 81,
    fig. 163.
Panicled, Paniculate: arranged in panicles, or like a panicle.
Papery: of about the consistence of letter-paper.
Papilionaceous: butterfly-shaped ; applied to such a corolla as that of the Pea
    and the Locust-tree ; p. 105, fig. 217.
Papílla (plural *papillæ*) : little nipple-shaped protuberances.
Papíllate, Papíllose: covered with papillæ.
Pappus: thistle-down. The down crowning the achenium of the Thistle, and
    other Compositæ, represents the calyx ; so the scales, teeth, chaff, as well
    as bristles, or whatever takes the place of the calyx in this family, are called
    the pappus ; fig. 292 – 296, p. 130.
Parallel-veined, or *nerved* (leaves) : p. 55, 56.
Paráphyses: jointed filaments mixed with the antheridia of Mosses. (Manual,
    p. 607.)
Parénchyma: soft cellular tissue of plants, like the green pulp of leaves.
Parietal (placentæ, &c.) : attached to the walls (*parietes*) of the ovary or peri-
    carp ; p. 119, 120.
Parted: separated or cleft into parts almost to the base ; p. 62.
Partial involucre, same as an *involucel: partial petiole,* a division of a main leaf-
    stalk or the stalk of a leaflet : *partial peduncle,* a branch of a peduncle : *par-
    tial umbel,* an umbellet, p. 81.
Patent: spreading ; open. *Patulous:* moderately spreading.
Pauci-, in composition : few ; as *pauciflorous,* few-flowered, &c.
Pear-shaped: solid obovate, the shape of a pear.
Pectinate: pinnatifid or pinnately divided into narrow and close divisions, like
    the teeth of a comb.
Pedate: like a bird's foot ; palmate or palmately cleft, with the side divisions
    again cleft, as in Viola pedata, &c.
Pedately cleft, lobed, &c. : cut in a pedate way.
Pédicel: the stalk of each particular flower of a cluster ; p. 78, fig. 156.
Pédicellate, Pédicelled: furnished with a pedicel.

Péduncle : a flower-stalk, whether of a single flower or of a flower-cluster ; p. 78.

Péduncled, Pedinculate : furnished with a peduncle.

Peltate : shield-shaped : said of a leaf, whatever its shape, when the petiole is attached to the lower side, somewhere within the margin ; p. 59, fig. 102, 178.

Pendent : hanging. *Pendulous :* somewhat hanging or drooping.

Penicillate : tipped with a tuft of fine hairs, like a painter's pencil ; as the stigmas of some Grasses.

Penta- (in words of Greek composition) : five ; as

Pentágynous : with five pistils or styles ; p. 116.

Pentámerous : with its parts in fives, or on the plan of five.

Pentándrous : having five stamens ; p. 112. *Pentástichous :* in five ranks.

Pepo : a fruit like the Melon and Cucumber; p. 128.

Perennial : lasting from year to year ; p. 21.

Perfect (flower) : having both stamens and pistils ; p. 89.

Perfóliate : passing through the leaf, in appearance ; p. 67, fig. 131, 132.

Pérforate : pierced with holes, or with transparent dots resembling holes, as an Orange-leaf.

Périanth : the leaves of the flower generally, especially when we cannot readily distinguish them into calyx and corolla ; p. 85.

Péricarp : the ripened ovary ; the walls of the fruit ; p. 127.

Pericárpic : belonging to the pericarp.

Périchœth : the cluster of peculiar leaves at the base of the fruit-stalk of Mosses.

Perichœtial : belonging to the perichœth.

Perigónium, Perigóne : same as *perianth.*

Perigýnium : bodies around the pistil ; applied to the closed cup or bottle-shaped body which encloses the ovary of Sedges, and to the bristles, little scales, &c. of the flowers of some other Cyperaceæ.

Perígynous : the petals and stamens borne on the calyx ; p. 104, 111.

Peripheric : around the outside, or periphery, of any organ.

Périsperm : a name for the albumen of a seed (p. 136).

Péristome : the fringe of teeth, &c. around the orifice of the capsule of Mosses. (Manual, p. 607.)

Persistent : remaining beyond the period when such parts commonly fall, as the leaves of evergreens, and the calyx, &c. of such flowers as remain during the growth of the fruit.

Pérsonate : masked ; a bilabiate corolla with a projection, or *palate,* in the throat, as of the Snapdragon ; p. 106, fig. 210, 211.

Petal : a leaf of the corolla ; p. 85.

Petaloid : petal-like ; resembling or colored like petals.

Pétiole : a footstalk of a leaf ; a leaf-stalk, p. 54.

Petioled, Petiolate : furnished with a petiole.

Petiólulate : said of a leaflet when raised on its own partial leafstalk.

Phœnógamous, or *Phanerógamous :* plants bearing flowers and producing seeds : same as Flowering Plants ; p. 177, 182.

Phyllódium (plural *phyllodia*) : a leaf where the blade is a dilated petiole, as in New Holland Acacias ; p. 69.

Phyllotáxis, or *Phyllotaxy :* the arrangement of leaves on the stem ; p. 71.

Physiological Botany, Physiology, p. 3.

Phyton : a name used to designate the pieces which by their repetition make up a plant, theoretically, viz. a joint of stem with its leaf or pair of leaves.

Piliferous : bearing a slender bristle or hair (*pilum*), or beset with hairs.

Pilose : hairy ; clothed with soft slender hairs.

Pinna : a primary branch of the petiole of a bipinnate or tripinnate leaf, as fig. 130, p. 66.

Pinnule : a secondary branch of the petiole of a bipinnate or tripinnate leaf ; p. 66.

Pinnate (leaf) : when the leaflets are arranged along the sides of a common petiole ; p. 65, fig. 126 – 128.

Pinnately lobed, cleft, parted, divided, &c., p. 63.

Pinnatifid : same as pinnately cleft ; p. 63, fig. 119.

Pistil : the seed-bearing organ of the flower ; p. 86, 116.

Pistillidium : the body which in Mosses, Liverworts, &c. answers to the pistil.

Pitchers, p. 51, fig. 79, 80.

Pith : the cellular centre of an exogenous stem ; p. 150, 151.

Pitted : having small depressions or pits on the surface, as many seeds.

Placenta : the surface or part of the ovary to which the ovules are attached ; p. 118.

Plaited (in the bud) ; p. 76, fig. 150 ; p. 110, fig. 225.

Plane : flat, outspread.

Plicate : same as *plaited.*

Plumose : feathery ; when any slender body (such as a bristle of a pappus) is beset with hairs along its sides, like the plumes or the beard on a feather.

Plumule : the little bud or first shoot of a germinating plantlet above the cotyledons ; p. 6, fig. 5 ; p. 137.

Pluri-, in composition : many or several ; as

Plurifoliolate : with several leaflets ; p. 66.

Pod : specially a legume, p. 131 ; also applied to any sort of capsule.

Podosperm : the stalk of a seed.

Pointless : destitute of any pointed tip, such as a *mucro, awn, acumination,* &c.

Pollen : the fertilizing powder of the anther ; p. 86, 114.

Pollen-mass : applied to the pollen when the grains all cohere into a mass, as in Milkweed and Orchis.

Poly- (in compound words of Greek origin) : same as *multi-* in those of Latin origin, viz many ; as

Polyadelphous : having the stamens united by their filaments into several bundles ; p. 112.

Polyandrous : with numerous (more than 20) stamens (inserted on the receptacle) ; p. 112.

Polycotyledonous : having many (more than two) cotyledons, as Pines ; p. 17, 137, fig. 45, 46.

Polygamous : having some perfect and some separated flowers, on the same or on different individuals, as the Red Maple.

Polygonal : many-angled.

Polygynous : with many pistils or styles ; p. 116.

Polymerous : formed of many parts of each set.

Polymorphous : of several or varying forms.

Polypetalous : when the petals are distinct or separate (whether few or many) ; p. 103.

Polyphyllous: many-leaved ; formed of several distinct pieces, as the calyx of Sedum, fig. 168, Flax, fig. 174, &c.

Polysépalous: same as the last when applied to the calyx ; p. 103.

Polyspérmous: many-seeded.

Pome: the apple, pear, and similar fleshy fruits ; p. 128.

Porous: full of holes or pores.

Pouch: the silicle or short pod, as of Shepherd's Purse ; p. 133.

Præfloration: same as *æstivation;* p. 108.

Præfoliation: same as *vernation;* p. 75.

Præmórse: ending abruptly, as if bitten off.

Prickles: sharp elevations of the bark, coming off with it, as of the Rose; p. 39.

Prickly: bearing prickles, or sharp projections like them.

Prímine: the outer coat of the covering of the ovule ; p. 124.

Primórdial: earliest formed ; primordial leaves are the first after the cotyledons.

Prismátic: prism-shaped ; having three or more angles bounding flat or hollowed sides.

Process: any projection from the surface or edge of a body.

Procumbent: trailing on the ground ; p. 37.

Produced: extended or projecting, as the upper sepal of a Larkspur is *produced* above into a spur ; p. 91, fig. 183.

Prolíferous (literally, bearing offspring) · where a new branch rises from an older one, or one head or cluster of flowers out of another, as in Filago Germanica, &c.

Prostrate: lying flat on the ground.

Próteine: a vegetable product containing nitrogen ; p. 165.

Prótoplasm: the soft nitrogenous lining or contents of cells, p. 165.

Prúinose, Pruinate: frosted ; covered with a powder like hoar-frost.

Pubérulent: covered with fine and short, almost imperceptible down.

Pubéscent: hairy or downy, especially with fine and soft hairs or *pubescence.*

Pulvérulent, or *Pulveraceous:* dusted ; covered with fine powder, or what looks like such.

Púlvinate: cushioned, or shaped like a cushion.

Punctate: dotted, either with minute holes or what look as such (as the leaves of St John's-wort and the Orange), or with minute projecting dots.

Pungent: very hard, and sharp-pointed ; prickly-pointed.

Putámen: the stone of a drupe, or the shell of a nut ; p. 128.

Pyramidal: shaped like a pyramid.

Pyréne, Pyréna: a seed-like nutlet or stone of a small drupe.

Pyxis, Pyxídium: a pod opening round horizontally by a lid ; p. 133, fig. 298, 311.

Quadri-, in words of Latin origin : four ; as

Quadrángular: four-angled   *Quadrifóliate:* four-leaved.

Quádrifid: four-cleft ; p 62.

Quatérnate · in fours.   *Quinate:* in fives.

Quincúncial: in a quincunx ; when the parts in æstivation are five, two of them outside, two inside, and one half out and half in, as shown in the calyx, fig. 224.

Quíntuple: five-fold.

Race: a marked variety which may be perpetuated from seed ; p. 174.

Raceme: a flower-cluster, with one-flowered pedicels arranged along the sides of a general peduncle ; p. 78, fig. 156.

Racemose: bearing racemes, or raceme-like.

Rachis: see *rhachis.*

Radial: belonging to the ray.

Radiate, or *Radiant:* furnished with ray-flowers ; p. 107.

Radical: belonging to the root, or apparently coming from the root.

Radicant: rooting, taking root on or above the ground, like the stems of Trumpet-Creeper and Poison-Ivy.

Radicels: little roots or rootlets.

Radicle: the stem-part of the embryo, the lower end of which forms the root ; p. 6, fig. 4, &c. ; p. 137.

Rameal: belonging to a branch. *Ramose:* full of branches (*rami*).

Ramulose: full of branchlets (*ramuli*).

Raphe: see *rhaphe.*

Ray: the marginal flowers of a head (as of Coreopsis, p. 107, fig. 219) or cluster (as of Hydrangea, fig. 167), when different from the rest, especially when ligulate, and diverging (like rays or sunbeams) ; the branches of an umbel, which diverge from a centre ; p. 79.

Receptacle: the axis or support of a flower ; p. 86, 124 ; the common axis or support of a head of flowers ; fig. 230.

Reclined: turned or curved downwards ; nearly recumbent.

Recurved: curved outwards or backwards.

Reduplicate (in æstivation) : valvate with the margins turned outwards, p. 109.

Reflexed: bent outwards or backwards.

Refracted: bent suddenly, so as to appear broken at the bend.

Regular: all the parts similar ; p. 89.

Reniform: kidney-shaped ; p. 58, fig. 100.

Repand: wavy-margined ; p. 62, fig. 115.

Repent: creeping, i. e. prostrate and rooting underneath.

Replum: the persistent frame of some pods (as of Prickly Poppy and Cress), after the valves fall away.

Reproduction, organs of: all that pertains to the flower and fruit ; p. 76.

Resupinate: inverted, or appearing as if upside down, or reversed.

Reticulated: the veins forming network, as in fig. 50, 83.

Retroflexed: bent backwards ; same as *reflexed.*

Retuse: blunted ; the apex not only obtuse, but somewhat indented ; p. 60, fig. 107.

Revolute: rolled backwards, as the margins of many leaves ; p. 76.

Rhachis (the backbone) : the axis of a spike, or other body ; p. 78.

Rhaphe: the continuation of the seed-stalk along the side of an anatropous ovule (p. 123) or seed ; fig. 273, *r,* 319 and 320, *b.*

Rhaphides: crystals, especially needle-shaped ones, in the tissues of plants.

Rhizoma: a rootstock , p. 40, fig. 64 - 67.

Rhombic: in the shape of a rhomb. *Rhomboidal:* approaching that shape.

Rib: the principal piece, or one of the principal pieces, of the framework of a leaf, p. 55 ; or any similar elevated line along a body.

Ring: an elastic band on the spore-cases of Ferns. (Manual, p. 587, plate 9, fig. 2, 3.)

Ringent: grinning; gaping open; p. 102, fig. 209.

Root, p. 28.

Root-hairs, p. 31, 149.

Rootlets: small roots, or root-branches ; p. 29.

Rootstock: root-like trunks or portions of stems on or under ground ; p. 40.

Rosaceous: arranged like the petals of a rose.

Rostellate: bearing a small beak (*rostellum*).

Rostrate: bearing a beak (*rostrum*) or a prolonged appendage.

Rosulate: in a regular cluster of spreading leaves, resembling a full or double rose, as the leaves of Houseleek, &c.

Rotate: wheel-shaped : p. 101, fig. 204, 205.

Rotund: rounded or roundish in outline.

Rudimentary: imperfectly developed, or in an early state of development.

Rugose: wrinkled, roughened with wrinkles.

Ruminated (albumen) : penetrated with irregular channels or portions filled with softer matter, as a nutmeg.

Runcinate: coarsely saw-toothed or cut, the pointed teeth turned towards the base of the leaf, as the leaf of a Dandelion.

Runner: a slender and prostrate branch, rooting at the end, or at the joints, as of a Strawberry, p. 38.

Sac: any closed membrane, or a deep purse-shaped cavity.

Sagittate: arrowhead-shaped ; p 59, fig. 95.

Salver-shaped, or *Salver-form*: with a border spreading at right angles to a slender tube, as the corolla of Phlox, p. 101, fig. 208, 202.

Samara: a wing-fruit, or key, as of Maple, p. 5, fig. 1, Ash, p. 131, fig. 300, and Elm, fig. 301.

Samaroid: like a samara or key-fruit.

Sap: the juices of plants generally. Ascending or crude sap; p. 161, 168. Elaborated sap, that which has been digested or assimilated by the plant ; p. 162, 169.

Sarcocarp: the fleshy part of a stone-fruit, p 128.

Sarmentaceous: bearing long and flexible twigs (*sarments*), either spreading or procumbent.

Saw-toothed : see *serrate*.

Scabrous: rough or harsh to the touch.

Scalariform: with cross-bands, resembling the steps of a ladder.

Scales: of buds, p. 22, 50 ; of bulbs, &c., p. 40, 46, 50.

Scaly: furnished with scales, or scale-like in texture ; p. 46, &c.

Scandent: climbing ; p. 37.

Scape: a peduncle rising from the ground, or near it, as of the stemless Violets, the Bloodroot, &c.

Scapiform: scape-like.

Scar of the seed, p. 135. *Leaf-scars*, p. 21.

Scarious or *Scariose*: thin, dry, and membranous.

Scobiform: resembling sawdust.

Scórpioid or *Scorpioidal :* curved or circinate at the end, like the tail of a scorpion, as the inflorescence of Heliotrope.

Scrobículate : pitted ; excavated into shallow pits.

Scurf, Scurfiness : minute scales on the surface of many leaves, as of Goosefoot, Buffalo-berry, &c.

Scútate : buckler-shaped.

Scutéllate, or *Scutélliform :* saucer-shaped or platter-shaped.

Sécund : one-sided ; i. e. where flowers, leaves, &c. are all turned to one side.

Secúndine : the inner coat of the ovule ; p. 124.

Seed, p. 134. *Seed-coats,* p. 134. *Seed-vessel,* p. 127.

Segment : a subdivision or lobe of any cleft body.

Séyregate : separated from each other.

Semi- (in compound words of Latin origin) : half; as

Semi-adherent, as the calyx or ovary of Purslane, fig. 214. *Semicordate :* half-heart-shaped. *Semilunar :* like a half-moon. *Semiovate :* half-ovate, &c.

Seminal : relating to the seed. *Seminiferous :* seed-bearing.

Sempérvirent : evergreen.

Sepal : a leaf or division of the calyx ; p. 85.

Sépaloid : sepal-like. *Sepaline :* relating to the sepals.

Separated Flowers : those having stamens or pistils only ; p. 89.

Septate : divided by partitions (*septa*):

Séptenate : with parts in sevens.

Septicídal : where a pod in dehiscence splits through the partitions, dividing each into two layers ; p. 132, fig. 306.

Septiferous : bearing the partition.

Septifragal : where the valves of a pod in dehiscence break away from the partitions ; p. 132.

Septum (plural *septa*) : a partition, as of a pod, &c.

Sérial, or *Seriate :* in rows ; as *biserial,* in two rows, &c.

Seríceous : silky ; clothed with satiny pubescence.

Serótinous : happening late in the season.

Serrate, or *Serrated :* the margin cut into teeth (*serratures*) pointing forwards ; p. 61, fig. 112.

Sérrulate : same as the last, but with fine teeth.

Sessile : sitting ; without any stalk, as a leaf destitute of petiole, or an anther destitute of filament.

Seta : a bristle, or a slender body or appendage resembling a bristle.

Setáceous : bristle-like. *Setiform :* bristle-shaped.

Setígerous : bearing bristles. *Setose :* beset with bristles or bristly hairs.

Sex : six ; in composition. *Sexangular :* six-angled, &c.

Sheath : the base of such leaves as those of Grasses, which are

Sheathing : wrapped round the stem.

Shield-shaped : same as *scutate,* or as *peltate,* p. 59.

Shrub, p. 21.

Sigmoid : curved in two directions, like the letter S, or the Greek *sigma.*

Silículose : bearing a silicle, or a fruit resembling it.

Sílicle : a pouch, or short pod of the Cress Family ; p. 133.

Silíque : a longer pod of the Cress Family ; p. 133, fig. 310.

Siliquose : bearing siliques or pods which resemble siliques.

Silky : glossy with a coat of fine and soft, close-pressed, straight hairs.

Silver-grain of wood , p. 151.

Silvery : shining white or bluish-gray, usually from a silky pubescence.

Simple : of one piece ; opposed to *compound*.

Sinistrorse : turned to the left.

Sinuate : strongly wavy ; with the margin alternately bowed inwards and outwards ; p. 62, fig. 116.

Sinus : a recess or bay ; the re-entering angle or space between two lobes or projections.

Sleep of Plants (so called), p. 170.

Soboliferous : bearing shoots from near the ground.

Solitary : single ; not associated with others.

Sorus (plural *sori*) : the proper name of a fruit-dot of Ferns.

Spadix : a fleshy spike of flowers ; p. 80, fig. 162.

Spathaceous : resembling or furnished with a

Spathe : a bract which inwraps an inflorescence ; p. 80, fig. 162.

Spatulate, or *Spathulate* : shaped like a spatula ; p. 58, fig. 92.

Special Movements, p. 170.

Species, p. 173.

Specific Character, p. 181.   *Specific Names*, p. 179.

Spicate : belonging to or disposed in a spike.

Spiciform : in shape resembling a spike.

Spike : an inflorescence like a raceme, only the flowers are sessile ; p. 80, fig. 160.

Spikelet : a small or a secondary spike ; the inflorescence of Grasses.

Spine : a thorn ; p. 39.

Spindle-shaped · tapering to each end, like a radish ; p. 31, fig. 59.

Spinescent : tipped by or degenerating into a thorn.

Spinose, or *Spiniferous* : thorny.

Spiral arrangement of leaves, p. 72.   *Spiral vessels or ducts*, p. 148.

Sporangia, or *Sporocarps* : spore-cases of Ferns, Mosses, &c.

Spore : a body resulting from the fructification of Cryptogamous plants, in them taking the place of a seed.

Sporule : same as a spore, or a small spore.

Spur : any projecting appendage of the flower, looking like a spur, as that of Larkspur, fig. 183.

Squamate, *Squamose*, or *Squamaceous* : furnished with scales (*squamæ*).

Squamellate or *Squamulose* : furnished with little scales (*squamellæ* or *squamulæ*).

Squamiform : shaped like a scale.

Squarrose : where scales, leaves, or any appendages, are spreading widely from the axis on which they are thickly set.

Squarrulose : diminutive of *squarrose* ; slightly squarrose.

Stalk : the stem, petiole, peduncle, &c., as the case may be.

Stamen, p. 86, 111.

Staminate : furnished with stamens ; p. 89.   *Stamineal* : relating to the stamens

Staminodium : an abortive stamen, or other body resembling a sterile stamen.

Standard : the upper petal of a papilionaceous corolla ; p. 105, fig. 217, 218, &

Starch : a well-known vegetable product ; p. 163.

Station: the particular place, or kind of situation, in which a plant naturally occurs.

Stéllate, Stéllular: starry or star-like; where several similar parts spread out from a common centre, like a star.

Stem, p. 36, &c.

Stemless: destitute or apparently destitute of stem.

Sterile: barren or imperfect; p. 89.

Stigma: the part of the pistil which receives the pollen; p. 87.

Stigmátic, or *Stigmatose:* belonging to the stigma.

Stipe (Latin *stipes*) · the stalk of a pistil, &c., when it has any; the stem of a Mushroom.

Stipel: a stipule of a leaflet, as of the Bean, &c.

Stipéllate: furnished with stipels, as the Bean and some other Leguminous plants.

Stipitate: furnished with a stipe, as the pistil of Cleome, fig. 276.

Stípulate: furnished with stipules.

Stipules: the appendages one each side of the base of certain leaves; p. 69.

Stolons: trailing or reclined and rooting shoots; p. 37.

Stoloníferous: producing stolons.

Stomate (Latin *stoma,* plural *stomata*): the breathing-pores of leaves, &c.; p. 156.

Strap-shaped: long, flat, and narrow; p. 106.

Striate, or *Striated:* marked with slender longitudinal grooves or channels (Latin *striæ*).

Strict: close and narrow; straight and narrow.

Strigíllose, Stríyose: beset with stout and appressed, scale-like or rigid bristles.

Strobildceous: relating to, or resembling a

Strôbile: a multiple fruit in the form of a cone or head, as that of the Hop and of the Pine; fig. 314, p. 133.

Strôphiole: same as *caruncle.* *Strophiolate:* furnished with a strophiole.

Struma: a wen; a swelling or protuberance of any organ.

Style: a part of the pistil which bears the stigma; p. 86.

Stylopôdium: an epigynous disk, or an enlargement at the base of the style, found in Umbelliferous and some other plants.

Sub-, as a prefix: about, nearly, somewhat; as *subcordate,* slightly cordate: *subserrate,* slightly serrate: *subaxillary,* just beneath the axil, &c., &c.

Súberose: corky or cork-like in texture.

Subclass, p. 177, 183. *Suborder,* p. 176. *Subtribe,* p. 177.

Súbulate: awl-shaped; tapering from a broadish or thickish base to a sharp point; p. 68.

Succulent: juicy or pulpy.

Suckers: shoots from subterranean branches; p. 37.

Suffrutéscent: slightly shrubby or woody at the base only; p. 36.

Sugar, p. 163.

Sulcate: grooved longitudinally with deep furrows.

Supernumerary Buds: p. 26.

Supérvolute: plaited and convolute in bud; p. 110, fig. 225.

Supra-axillary: borne above the axil, as some buds; p. 26, fig. 52.

Supra-decompound: many times compounded or divided.

Súrculose: producing suckers, or shoots resembling them.

Suspended: hanging down. Suspended ovules or seeds hang from the very summit of the cell which contains them; p. 122, fig. 269.

Sútural: belonging or relating to a suture.

Súture: the line of junction of contiguous parts grown together; p. 117.

Sword-shaped: vertical leaves with acute parallel edges, tapering above to a point; as those of Iris, fig. 133.

Symmetrical Flower: similar in the number of parts of each set; p. 89.

Syndntherous, or *Syngenesious:* where stamens are united by their anthers; p. 112, fig. 229.

Syncárpous (fruit or pistil): composed of several carpels consolidated into one.

System. p. 195.

Systematic Botany: the study of plants after their kinds; p. 3.

Taper-pointed: same as *acuminate;* p. 60, fig. 103.

Tap-root: a root with a stout tapering body; p. 32.

Tawny: dull yellowish, with a tinge of brown.

Taxónomy: the part of Botany which treats of classification.

Tégmen: a name for the inner seed-coat.

Tendril: a thread-shaped body used for climbing, p. 38: it is either a branch, as in Virginia Creeper, fig. 62; or a part of a leaf, as in Pea and Vetch, fig. 127.

Térete: long and round; same as *cylindrical,* only it may taper.

Términal: borne at, or belonging to, the extremity or summit.

Terminólogy: the part of the science which treats of technical terms; same as *glossology.*

Térnate: in threes; p. 66. *Ternately:* in a ternate way.

Testa: the outer (and usually the harder) coat or shell of the seed; p. 134.

Tetra- (in words of Greek composition): four; as,

Tetracóccous: of four cocci or carpels.

Tetradýnamous: where a flower has six stamens, two of them shorter than the other four, as in Mustard, p. 92, 112, fig. 188.

Tetrágonal: four-angled. *Tetrágynous:* with four pistils or styles; p. 116.

Tetrámerous: with its parts or sets in fours.

Tetrándrous: with four stamens; p. 112.

Theca: a case; the cells or lobes of the anther.

Thorn: see *spine;* p. 39.

Thread-shaped: slender and round, or roundish like a thread; as the filament of stamens generally.

Throat: the opening or gorge of a monopetalous corolla, &c., where the border and the tube join, and a little below.

Thyrse or *Thyrsus:* a compact and pyramidal panicle; p. 81.

Tómentose: clothed with matted woolly hairs (*tomentum*).

Tongue-shaped: long, flat, but thickish, and blunt.

Toothed: furnished with teeth or short projections of any sort on the margin, used especially when these are sharp, like saw-teeth, and do not point forwards; p. 61, fig. 113.

Top-shaped: shaped like a top, or a cone with its apex downwards.

Tórose, Tórulose: knobby; where a cylindrical body is swollen at intervals.

Torus: the receptacle of the flower; p. 86, 124.

Tree, p. 21.

Tri-, in composition: three; as

Triadélphous: stamens united by their filaments into three bundles; p. 112.

Tridndrous: where the flower has three stamens; p. 112.

Tribe, p. 176.

Trichótomous: three-forked. *Tricóccous:* of three cocci or roundish carpels.

Tricolor: having three colors. *Tricóstate:* having three ribs.

Tricúspidate: three-pointed. *Tridéntate:* three-toothed.

Triénnial: lasting for three years.

Trifárious: in three vertical rows; looking three ways.

Trífid: three-cleft; p. 62.

Trifóliate: three-leaved. *Trifóliolate:* of three leaflets; p. 66.

Trifúrcate: three-forked. *Trígonous:* three-angled, or triangular.

Trígynous: with three pistils or styles; p. 116. *Tríjugate:* in three pairs (*jugi*).

Trilóbed, or *Trilobate:* three-lobed; p. 62.

Trilócular: three-celled, as the pistils or pods in fig. 225 – 227.

Trímerous: with its parts in threes, as Trillium, fig. 189.

Trinérvate: three-nerved, or with three slender ribs.

Triócious: where there are three sorts of flowers on the same or different individuals; as in Red Maple.

Tripártible: separable into three pieces. *Tripártite:* three-parted; p. 62.

Tripetalous: having three petals; as in fig. 189.

Triphýllous: three-leaved; composed of three pieces.

Tripínnate: thrice pinnate; p. 66. *Tripinnátifid:* thrice pinnately cleft; p. 64.

Triple-ribbed, Triple-nerved, &c.: where a midrib branches into three near the base of the leaf, as in Sunflower.

Triquétrous: sharply three-angled; and especially with the sides concave, like a bayonet.

Trisérial, or *Triseriate:* in three rows, under each other.

Trístichous: in three longitudinal or perpendicular ranks.

Tristigmátic, or *Tristígmatose:* having three stigmas.

Trisúlcate: three-grooved.

Tritérnate: three times ternate; p. 67.

Trivial Name: the specific name.

Trochlear: pulley-shaped.

Trumpet-shaped: tubular, enlarged at or towards the summit, as the corolla of Trumpet-Creeper.

Truncate: as if cut off at the top; p. 60, fig. 106.

Tube, p. 102.

Trunk: the main stem or general body of a stem or tree.

Tuber: a thickened portion of a subterranean stem or branch, provided with eyes (buds) on the sides; as a potato, p. 43, fig. 68.

Túbercle: a small excrescence.

Tubercled, or *Tuberculate:* bearing excrescences or pimples.

Túberous: resembling a tuber. *Tuberíferous:* bearing tubers.

Túbular: hollow and of an elongated form; hollowed like a pipe.

Tumid: swollen ; somewhat inflated.

Túnicate: coated ; invested with layers, as an onion ; p. 46.

Túrbinate: top-shaped. *Turgid:* thick as if swollen.

Túrio (plural *turiones*) : young shoots or suckers springing out of the ground ; as Asparagus-shoots.

Turnip-shaped: broader than high,   narrowed below ; p. 32, fig. 57.

Twin: in pairs (see *gemiuate*), as the flowers of Linnœa

Twining: ascending by coiling round a support, like the Hop ; p. 37.

Typical: well expressing the characteristics of a species, genus, &c.

Úmbel: the umbrella-like form of inflorescence ; p. 79, fig. 159.

Umbéllate: in umbels. *Umbellíferous:* bearing umbels.

Úmbellet: a secondary or partial umbel ; p. 81.

Umbílicate: depressed in the centre, like the ends of an apple.

Úmbonate: bossed ; furnished with a low, rounded projection like a boss (*umbo*)·

Umbráculiform: umbrella-shaped, like a Mushroom, or the top of the style of Sarracenia.

Unarmed: destitute of spines, prickles, and the like.

Úncinate: hook-shaped ; hooked over at the end.

Under-shrub: partially shrubby, or a very low shrub.

Úndulate: wavy, or wavy-margined ; p. 62.

Unequally pinnate: pinnate with an odd number of leaflets ; p. 65.

Unguículate: furnished with a claw (*unguis*) ; p. 102, i. e. a narrow base, as the petals of a Rose, where the claw is very short, and those of Pinks (fig. 200), where the claw is very long.

Uni-, in compound words : one ; as

Uniflórous: one-flowered. *Unifóliate:* one-leaved.

Unifóliolate: of one leaflet ; p. 66. *Unijugate:* of one pair.

Unildbiate: one-lipped. *Unilateral:* one-sided.

Unilócular: one-celled, as the pistil in fig. 261, and the anther in fig. 238, 239.

Uniovulate: having only one ovule, as in fig. 213, and fig. 267 – 269.

Unisérial: in one horizontal row.

Unisexual: having stamens or pistils only, as in Moonseed, fig. 176, 177, &c.

Únivalved: a pod of only one piece after dehiscence, as fig. 253.

Urceólate: urn-shaped.

Útricle: a small, thin-walled, one-seeded fruit, as of Goosefoot ; p. 130, fig. 350.

Útrícular: like a small bladder.

Váginate: sheathed, surrounded by a sheath (*vagina*).

Valve: one of the pieces (or doors) into which a dehiscent pod, or any similar body, splits ; p. 131, 114.

Valvate, Válvular: opening by valves. *Valvate* in æstivation, p. 109.

Variety, p. 174, 177.

Váscular: containing vessels, or consisting of vessels, such as ducts ; p. 146, 148.

Vaulted: arched ; same as *fornicate.*

Vegetable Physiology, p. 3.

Veil: the calyptra of Mosses. (Manual, p. 607)

Veins: the small ribs or branches of the framework of leaves, &c. ; p. 55.

Veined, Veiny: furnished with evident veins.　*Veinless:* destitute of veins.

Veinlets: the smaller ramifications of veins.

Velate: furnished with a veil.

Velútinous: velvety to the touch.

Venation: the veining of leaves, &c. ; p. 55.

Venose: veiny ; furnished with conspicuous veins.

Ventral: belonging to that side of a simple pistil, or other organ, which looks
　　towards the axis or centre of the flower ; the opposite of dorsal ; as the
Ventral Suture, p. 117.

Ventricose: inflated or swelled out on one side.

Venulose: furnished with veinlets.

Vermicular: shaped like worms.

Vernation: the arrangement of the leaves in the bud ; p. 75.

Vernicose: the surface appearing as if varnished.

Verrucose: warty ; beset with little projections like warts.

Versatile: attached by one point, so that it may swing to and fro, as the anthers
　　of the Lily and Evening Primrose ; p. 113, fig. 234.

Vertex: same as the *apex.*

Vertical: upright ; perpendicular to the horizon, lengthwise.

Verticil: a whorl ; p. 71.　*Verticillate:* whorled ; p. 71, 75, fig. 148.

Vesicle: a little bladder.　*Embryonal Vesicle,* p. 139.　*Vesicular:* bladdery.

Vessels: ducts, &c. ; p. 146, 148.

Vexillary, Vexillar: relating to the

Vexillum: the standard of a papilionaceous flower ; p. 105, fig. 218, *s.*

Villose: shaggy with long and soft hairs (*villosity.*)

Viuímeous: producing slender twigs, such as those used for wicker-work.

Vine: any trailing or climbing stem ; as a Grape-vine.

Virescent, Viridescent: greenish ; turning green.

Virgate: wand-shaped, as a long, straight, and slender twig.

Viscous, Viscid: having a glutinous surface.

Vitta (plural *vittæ*) : the oil-tubes of the fruit of Umbelliferæ.

Voluble: twining, as the stem of Hops and Beans ; p. 37.

Wavy: the surface or margin alternately convex and concave ; p. 62.

Waxy: resembling beeswax in texture or appearance.

Wedge-shaped: broad above, and tapering by straight lines to a narrow base,
　　p. 58, fig. 94.

Wheel-shaped: see *rotate;* p. 102, fig. 204, 205.

Whorl, Whorled: when leaves, &c. are arranged in a circle round the stem,
　　p. 71, 75, fig. 148.

Wing: any membranous expansion.　*Wings* of papilionaceous flowers, p. 105.

Winged: furnished with a wing ; as the fruit of Ash and Elm, fig. 300, 301.

Wood, p. 145.　*Woody:* of the texture or consisting of wood.

Woody Fibre, or *Wood-Cells,* p. 146.

Woolly: clothed with long and entangled soft hairs ; as the leaves of Mullein.

THE END.

GRAY'S
BOTANIST'S MICROSCOPE.

*This Convenient Instrument, devised and manu-
factured first for the use of the Students in*

HARVARD UNIVERSITY,

*has given so great satisfaction there, and elsewhere,
that we deem it a duty to make it better known,
and offer it at a price within the reach of all
students.*

It is attached to a box, one and a half inches high and less than four inches long, into which it is neatly folded when not in use. The needles are used for dissecting flowers, or other objects, too small to be otherwise handled for analysis. The lenses magnify about **fifteen** diameters; or, with three lenses, about one-third more.

A thousand things about forest, field or garden, afford objects of intense interest for daily study.

Prof. ASA GRAY, of Harvard University, our popular American Botanist, says of it: "You are at liberty to call it the "GRAY'S MICROSCOPE." I do not think anything better can be made for the money."

Price of Microscope, with two lenses, - $2 00
"   "   " three " - 2 50